On Space and Time

ALAIN CONNES, MICHAEL HELLER,
SHAHN MAJID, ROGER PENROSE,
JOHN POLKINGHORNE AND
ANDREW TAYLOR

Edited by
SHAHN MAJID

CAMBRIDGE UNIVERSITY PRESS
Cambridge, New York, Melbourne, Madrid, Cape Town, Singapore, São Paulo, Delhi, Tokyo, Mexico City

Cambridge University Press
The Edinburgh Building, Cambridge CB2 8RU, UK

Published in the United States of America by Cambridge University Press, New York

www.cambridge.org
Information on this title: www.cambridge.org/9780521889261

First published 2008
Third printing 2011
Canto Classics edition 2012

Printed and bound in the United States of America

A catalogue record for this publication is available from the British Library

Library of Congress Cataloguing in Publication data
On space and time / Alain Connes . . . [et al.] ; edited by Shahn Majid.
p. cm.
Includes index.
ISBN 978-0-521-88926-1 (hardback)
1. Space and time. I. Connes, Alain. II. Majid, Shahn. III. Title.
QC173.59.S65O49 2008
523.1 – dc22 2008013428

ISBN 978-0-521-88926-1 Hardback

'Absolute, true, and mathematical time, of itself, and from its own nature, flows equably without relation to anything external'

Isaac Newton

'Already the distance-concept is logically arbitrary; there need be no things that correspond to it, even approximately'

Albert Einstein

'The future of our space and time / Is not gonna wither and die / The future of our space and time / Is not gonna say good-bye'

Vanessa Paradis/Lenny Kravitz

Contents

About the authors

Alain Connes holds the chair in Analysis and Geometry at the College de France and is Professor at the IHES in Paris and at Vanderbilt University in the USA. Awarded a Fields Medal in 1982 and the Crafoord Prize in 2002, he has pioneered the field of noncommutative geometry and its diverse applications in pure mathematics and physics. He is author of a major textbook in the field and numerous research articles.

Michael Heller is Professor of Philosophy at the Pontifical Academy of Theology in Cracow, Poland, and an adjunct member of the Vatican Observatory staff. An ordained Roman Catholic priest, he is author of numerous books and research articles on cosmology, philosophy and theology.

Shahn Majid is Professor of Mathematics at Queen Mary, University of London. Trained as a theoretical physicist and mathematician at Cambridge and Harvard, he helped pioneer the theory of quantum symmetry in the 1980s and 1990s. He is author of two textbooks in the field and numerous research articles.

Sir Roger Penrose is Emeritus Rouse Ball Professor of Mathematics at the University of Oxford. Awarded the 1988 Wolf Prize jointly with Stephen Hawking, he discovered twistor theory in the 1970s and has numerous other results from tilings to astrophysics and quantum theory. Popular books such as *The Emperor's New Mind* have presented his ideas about the human mind and the relationships between mathematics and physics.

John Polkinghorne KBE is a renowned Anglican theologian and former president of Queens' College, Cambridge, with a previous career as a leading particle physicist at the University of Cambridge in the 1960s and 1970s. Awarded the Templeton Prize in 2002, he is author of numerous books and research articles on physics and on theology.

Andrew Taylor is Professor of Astrophysics at the University of Edinburgh. He has made many major contributions to the field of cosmology, especially imaging the distribution of dark matter, studying the nature of dark energy and investigating the initial conditions of the Universe. He lives in Edinburgh with his wife and son.

Preface

What can more than two thousand years of human thought and several hundred years of hard science tell us finally about the true nature of space and time? This is the question that the philosopher Jeremy Butterfield and I posed to a unique panel of top mathematicians, physicists and theologians in a public discussion that took place at Emmanuel College, Cambridge in September 2006, and this is the book that grew out of that event. All four other panellists, myself and the astronomer Andy Taylor who spoke at a related workshop, now present our personal but passionately held insights in rather more depth.

The first thing that can be said is we do not honestly know the true nature of space and time. Therefore this book is not about selling a particular theory but rather it provides six refreshingly diverse points of view from which the reader can see that the topic is very much as alive today as it was at the time of St Augustine or of Newton or of Einstein. Our goal is not only to expose to the modern public revolutionary ideas at the cutting edge of theoretical physics, mathematics and astronomy but to expose that there is a debate to be had in the first place and at all levels, including a wider human context. Moreover, the reader will find here essays from leading figures unconstrained by peer pressure, fashion or dogma. These are views formed not today or yesterday but in each case over a scientific lifetime of study. So what I think this volume offers is six particularly coherent visions of lasting value at a time when serious thought on this topic is in a state of flux.

My own view as editor, and I think this comes across in all of the essays, is that right now is an enormously exciting juncture for fundamental science. There is real adventure to be had at a time in which pure mathematics, theoretical physics, astronomy, philosophy and experiment are all coming together in a manner unseen for almost a hundred years. You probably have to go back still further to the seventeenth-century Scientific Revolution to get a full sense of what I mean, to the era of 'natural philosophy', Copernicus, Galileo, Newton. I do sometimes hear some of my physicist colleagues lamenting the old days when physics was young, when there was not a vast mountain of theory or technique that a young researcher had to climb before even starting out. What I hope the essays in this volume will convey is a sense that actually *we are at the birth of a new such era right now.*

To explain this further I need first to explain where physics has been 'stuck' for so long. The sticking point – which is also the reason that no physicist can honestly say at the moment that they truly understand space and time – is what I call in my own essay 'a hole in the heart of science' and which the other essays refer to variously as 'the Planck scale' or 'the problem of quantum gravity'. Perhaps many readers will know that Einstein in his 1915 'General Theory of Relativity' provided what is still the current framework for how we think about gravity, as a curved spacetime. It is a theory that governs the large scale and among its subsequent solutions was the remarkable Big-Bang model of an expanding Universe which, with variations, is still used today. Einstein's earlier work also had some input into a theory of quantum mechanics which emerged in 1923, a theory which he, however, never fully accepted. Quantum mechanics evolved in the 1960s to its modern form 'quantum field theory' and this is our current best understanding of physics at the small scale of subatomic particles – quarks, electrons, neutrinos, and so forth – the resulting Standard Model well describes such matter *in* space and time as well as all fundamental forces other than gravity. But what still eludes us since those days is that these two parts of physics do not even now form a single self-consistent theory that could if it was known be called 'quantum gravity'. Without a theory of quantum gravity a theoretical physicist cannot pretend to 'truly' understand space and time. Without a true understanding we cannot, for example, say with confidence what happens at the very centre of a black hole (and modern thinking is that many galaxies have a huge black hole in their core) and nor can we say anything with certainty about the origin of the Universe, the very start of the 'Big Bang'. Let me stress, it is not only that we do not know, *we do not even have a theory to test* about this deepest layer of physics.

How then can a research scientist get up in the morning and go about their business if we do not understand such basic notions as space and time? The answer lies in two aspects of the way that science is done, something which nonscientists do not always realise. The first is that every bit of scientific knowledge generally has a domain of validity attached to it and in which it is supposed to hold. So you generally say something precise and accurate but under certain approximations where the things that you do not understand are estimated to be negligible. This is how we can have today what is a vast and marvellous body of accumulated knowledge accurate to many decimal places even if we do not understand the things that are most basic. Secondly, even if you want to explore the domain where all you really know is that you don't really know, or even if you just want to push the boundaries a little, you do have to put forth concrete ideas, postulates or specific exploratory models if you are going to get anywhere, sometimes even knowing full well that the model is not quite right but just hoping to get a bit of insight. This is the *theoretical* side of physics, that is to say you put forth bold or less bold hypotheses and see where they get you. In this case science, particularly at the fundamental end, does not have the certainty that the image of scientists in white coats often elicits in the public. It is more like exploring empty blackness armed with nothing more than a flashlight and a measuring tape.

Now, whereas Einstein was not content to 'give up' on the problem and spent the last years of his life looking for such a unified theory, the mainstream physics establishment for the most part *did* give up in a certain sense. I do not mean that physicists stopped talking about quantum gravity but in many ways it became the elephant in the room that one perhaps pays lip-service to and then goes away and solves something more tractable. Either that or it became a play area for often interesting but ultimately wild and somewhat random speculation slightly outside the mainstream and certainly divorced from the cardinal rule of serious physics – testability. I do not mean either that young physicists coming up through the system did not and do not dream of finding a theory of quantum gravity, most of us cherish this dream from childhood and something like that does not go away so easily. But the conventional wisdom until very recently was that we had *neither* deep enough ideas for a theory *nor* the hope of ever being able to test a theory if we had one. A double whammy! For example, many physicists will admit over a beer or a glass of wine that part of the problem is that spacetime is probably not a continuum, but they have no mathematical alternative, so they carry on building their theory on a continuum assumption. I include here string theory, which seeks to encode particles and forces as a quantum theory of small bits of string – but still moving in a continuum. On the other problem, even a few years ago it was inconceivable that quantum gravity could ever be tested in the laboratory in the foreseeable future. This was a simple matter of back-of-envelope estimates that gravity is so weak a force that its effects at a subatomic scale are absolutely tiny. I recall even three years ago a government grant application being angrily rejected because it implied that a certain quantum-gravity experiment *could* be done, as this would clearly be a waste of resources. Now I think this has all fed in recent decades into a certain malaise. It is a malaise that occurs when a long-standing problem is never really addressed and when most people believe that it is not even worth sincerely to try. I am not saying that nothing worthwhile has been done in the last two decades, far from it, rather my point is to analyse why the problem of an actual theory of quantum gravity has been so intractable and I think it is for the two very good reasons mentioned. And perhaps it could be said that in such a climate mainstream theoretical physics has in recent years lost a certain freshness of purpose, or if you like what has been acquired in the last two decades is a certain tiredness.

As editor of this volume I will be very happy if this reminder that *we do not really know* comes across more openly. This is an antidote to what I feel in recent decades has been a tendency in the media to give just the opposite impression. When one of my colleagues goes on the radio and provides an authoritative soundbite that spacetime is, for example, a 10-dimensional continuum (of which we are somehow constrained to live in 4) they give such an opposite impression. The correct statement is not that a version of string theory predicts that spacetime is a 10-dimensional continuum but that string theory *presumes* as a starting point that spacetime is a continuum of some dimension n in which extended objects called 'strings' move. String theory then turns out not to work in the desired case $n = 4$ due to some technical anomalies but one can fix these by taking $n = 10$, say. This 'fix' of course opens up much worse conceptual and technical problems about explaining away the unobserved extra dimensions and why this particular fix and not some other. What is lost to the public here is a sense of perspective, that this or that theory is just that, a theory that should be tested or if that is not possible then at least weighed for its explanatory power against its complexity and ad-hocness of presumptions. I should say that I have used the example of string theory here only because it has been so much in the media in recent years, the same criteria should be applied elsewhere in theoretical physics.

Now let me say why I think the conventional wisdom above about the unreachability of quantum gravity was unnecessarily pessimistic, and what else has quietly been going on behind the scenes. In fact both of the fundamental problems mentioned are to do with a dearth of imagination. First of all on the experimental side astronomers and Earth-based experimental physicists have made fantastic strides in recent years. We now have a convincing picture of the large-scale structure of the Universe from an observational side that throws up real puzzles that lack even one convincing idea for a theoretical explanation. These include issues such as the nature of dark matter and dark energy or as some would say 'the problem of the cosmological constant'. We also have new experimental technology such as very long baseline laser interferometry that will allow us for the first time to see the far and hence early Universe through gravitational waves and to test out radical new ideas for its origin. At the moment there are such detectors as LIGO and VIRGO based on Earth and another in development, LISA, to be based in space. Such technology could also be adapted to test for various quantum-gravity effects. Data from active galactic nuclei, from gamma-ray bursts hopefully collected by NASA's GLAST satellite to be launched in the near future, neutrino oscillations, all of these theoretically could now test different ideas for quantum gravity. So for the first time in the history of mankind we *can* actually put theories of quantum gravity to some kind of test. Some of this is fortunate happenstance but much of it is lateral thinking about how the tiniest quantum-gravity effects could in some cases 'amplify' to a measurable level. And all of this is *in addition* to certain long-standing puzzles such as why some fundamental particles have mass while others are massless. It was recently discovered that enigmatic particles called neutrinos do in fact have a small but nonzero mass, which was a surprise. In fact the Standard Model of particle physics has some two dozen free parameters and we have little idea why they take the values that they do, why elementary particles seem to be organised into mysterious repeating families, and so forth. Secondly, certain long-simmering but radical mathematical tools and concepts outside mainstream physics have also recently reached a critical level of maturity. One of them is quite simply abstract algebra which two decades ago was of no real interest to physicists but which is now accepted as a rich vein of structure that can replace continuum geometry. My own field has played a role here. Another is twistor theory in which geometry is built not on points but on light rays. I will say more about these topics later in this preface. So we live in an era where our schoolchildren should know that our most fundamental concepts are up for grabs, but that there are real theoretical and conceptual puzzles and real experimental data coming online that can test new and creative ideas, that the Universe is a totally fascinating, mysterious and yet scientifically knowable place.

Up to a point. It still seems unlikely that current scientific considerations alone can in fact provide the final answer, but perhaps they are elements of some kind of emerging new renaissance exactly centred on our understanding of space and time. What is already clear is that the true problems of quantum gravity also require deep philosophical input about the nature of quantum measurement, the nature of time. I think they force us to think about what physical reality itself is more generally. And I do not see this as a one-way street in which scientists only inform the public. Scientists' ideas have to come from somewhere too, from sitting in cafés, from art, from life. Let us also not forget that many of our great universities have their roots in the middle ages founded out of theological institutions. The 1347 statues of my own former Cambridge college, Pembroke, as recorded in 1377, precisely list the main topics of study as grammar, logic and natural philosophy at the BA level, philosophy and some mathematics at the MA level, and these so-called 'arts' and theology at the more senior level. Later on science largely won out as the fount of physical knowledge but if science is now short of ideas on the deepest issues we should not rule out a thoughtful wider dialogue.

And so this is how this volume is put together. Each of the authors brings their own unique expertise to bear on the problem. Andrew Taylor's work on gravitational lensing has helped establish the large-scale structure of the Universe. He is also part of a consortium with recent breakthrough results on mapping out dark matter which is a key unknown at the moment. His essay provides a full account of this current experimental picture of the Universe at a cosmological scale and some of Taylor's personal insights into the weak spots and fundamental problems. The Fields medallist Alain Connes is perhaps at the other end of the spectrum, a pure mathematician of the highest level who has seen deeply into the nature of geometry itself and into how it could be reinvented in terms of operator algebras. He is an acknowledged pioneer and world leader of this new field of noncommutative geometry. In his essay he shows how his 'spectral' vision of geometry elegantly encodes the plethora of elementary particles found by physicists as a finite extra bit of such geometry tagged onto ordinary spacetime. It provides a new geometrical way of thinking about matter itself. Remarkably, his formulation also predicts relations between the masses of certain elementary particles as well as an estimate of the Higgs particle mass, all of which could be tested in principle. At the end of his essay Connes also speculates on connections between quantum gravity and number theory. My own essay comes to a kind of noncommutative geometry, but down a slightly different road, namely from my work on 'quantum symmetry' at the pure end of mathematical physics. Such considerations lead to models of 'quantum spacetime' where exact points in space and time fundamentally do not exist due to quantum 'fuzziness'. Moreover, gamma-ray bursts and a modification of technology such as LISA could in principle test the theory (predictions include a variation of the speed of light). If quantum spacetime was ever observed experimentally it would be a new fundamental effect in physics. This vision and that of Connes have common ground and a synthesis is certainly possible. In the later part of my essay I speculate on deeper philosophical ideas about the self-dual nature of physical reality in which quantum symmetries naturally fit. A very different vision appears in the essay of Sir Roger Penrose, inventor among other things of 'twistor theory' in the 1970s. This says that light rays and not points should be the fundamental ingredients in a true understanding of spacetime. Due to time dilation a photon of light does not itself experience time but exists simultaneously from its creation to its destruction. An important part of twistor theory is the idea of conformal invariance and this can be used for the 'conformal compactification' of spacetime (or the method of Penrose diagrams as they have been called). Penrose uses this to present a radical new proposal for what happened *before* the Big-Bang origin of the Universe. In this theory if you look back far enough you should see remnant information from a previous universe, the infinite future of which is the origin of ours (and similarly our infinite future is the origin of a next universe). Although it is hard at the moment to quantify, gravitational wave detectors such as LISA could play a role in looking for such effects. Observational input such as the cosmic microwave background also plays a role in the arguments, while most profound is perhaps an explanation of how such a cyclic (endlessly repeating) vision of cosmology is compatible with the thermodynamic arrow of time in which entropy always increases. Penrose argues here that one must include the entropy of the gravitational field to balance the accounts. His essay also outlines some of his other ideas about gravity, quantum theory and causality, exposed further in his several books. The volume is rounded off by the essays of Michael Heller, philosopher, theologian and cosmologist at the Vatican observatory, noted for discussions on scientific matters with Pope John Paul II, and of John Polkinghorne, former Cambridge Professor of Physics and a distinguished Anglican theologian. Heller's article particularly explores the philosophical and theological consequences of the quantum nonlocality suggested by our emerging understanding of quantum gravity. Conversely he explains how many of the notions of Western science have their deeper roots in history and that it would be better to be aware of these past influences than to leave them hidden and unquestioned. Polkinghorne in his short essay focusses specifically on the nature of time and contrasts the atemporal 'block' concept in which we look down on time from the outside, which was also Einstein's view, and the need for a different doctrine of 'temporal becoming' that really expresses how time unfolds.

Finally, let me say that both my own and the essay of the astronomer Andy Taylor have pedagogical sections where some of the background material needed in other essays is explained at a less technical level. Hence the reader could consider starting with some sections of these essays. By contrast, the visions of Connes and Penrose are probably the most profound but will also require more work from the reader fully to grasp. The essay of Heller could be read directly but with backward reference for some of the science, while the essay of Polkinghorne is both a very accessible epilogue to the entire volume and an introduction to his own two recent books. In all the works, because we are talking about fundamental physics, the Universe, etc., there will be references to vast or to vastly small numbers. The kinds of numbers involved are necessarily mind-boggling even for a physicist and we will have to use scientific notation for them. Here a large number is expressed in terms of 1 followed by so many zeros. For example an American billion is 10^9 meaning 1 000 000 000 (1 followed by nine zeros). Similarly 10^{-9} means one American billionth, i.e. start with 1.0 and then move the decimal place 9 spots to the *left* to obtain 0.000 000 001. In terms of physical units, cm means centimetres, g means grams, s means seconds and K means degrees Kelvin (degrees centigrade + 273.15). Other units will generally be explained where needed. We will generally state physical quantities only to a level of accuracy sufficient for the discussion.

The cover is from Corpus Christi College, Cambridge, 'New Court'. Warm thanks to Ian Fleming for doing the photography for this, and to the college for access. The background is galactic cluster CL0024+17 overlaid by a fog representing the apparent distribution of dark matter, courtesy of NASA, ESA, M. J. Jee and H. Ford *et al.* at Johns Hopkins University.

Shahn Majid, Cambridge 2007

1 The dark Universe

Andrew Taylor

'Space and Time are the modes by which we think, not the conditions in which we live' – *Albert Einstein*

'The only reason for time is so that everything doesn't happen at once' – *Albert Einstein*

'Time is an illusion. Lunchtime doubly so' – *Douglas Adams*

1.1 SPACE AND TIME IN COSMOLOGY

The question about the nature of space and time is intimately linked with the question of cosmology: Did space and time have a beginning? Do they go on forever? Space and time form the framework for our picture of cosmology, while our large-scale view of the Universe puts the limits on what space and time are.

The nature of space and time underwent a radical change from Newton to Einstein. As Newton set out in his Principia Mathematica, space and time was an unchanging Aristotelian background to the unfolding play of particles and waves. But even this seemingly innocuous assumption caused Newton problems. Gravity acted instantaneously everywhere (action at a distance); a radical idea for the 1770s used to the idea that every effect had a direct cause. If the Universe was infinite in extent, the forces acting on any given point would depend instantaneously on the influence of all of the matter throughout the Universe. But because the volume of space increases rapidly with distance these forces would accumulate and increase without limit in an infinite Universe. These problems were mainly swept under the carpet as Newtonian gravity clearly gave an excellent local approximation

On Space and Time, ed. Shahn Majid. Published by Cambridge University Press.

to the motion of the moon and planets. But these divergences made it clear that Newtonian gravity and Aristotelian space and time could not form the basis of a consistent cosmology.

When Einstein introduced his idea that gravity was a manifestation of curved space and time, allowing gravitational effects to be transmitted causally due to gravitational waves, these problems were removed. Not only was a given point in the Universe affected only by a causal region around it, but the whole of space and time became dependent on the distribution and abundance of matter residing in it.

The appearance of dynamical and wave-like properties of space and time opened up a new problem about its small-scale nature. In the Newtonian picture space and time were convenient and unchanging coordinates, detailing the changing positions of particles and waves. To ask what space and time were 'made of' in this scenario did not make much sense. Space and time are just a way to distinguish between different points and events, and had no extra existence. Even in the Einstein picture what is being described is the relationship between events and perhaps we still have no right to ask about the nature of space and time, what goes on in between the events. However, now that our coordinate system has taken on a more dynamical appearance, this does seem to suggest the space and time has much more structure to it than before and is therefore open to further investigation.

1.2 THE EXPANDING UNIVERSE

Perhaps the most impressive example of the dynamic nature of space and time in Einstein's theory of gravity is the expansion of the Universe. The discovery of the expansion of the Universe rightly rests with the American Astronomer Vesto Slipher (and not Edwin Hubble as commonly assumed) who, in a period between 1912 and 1920, noted that the light from the nebulae that he was observing seemed to be shifted to the red. He interpreted this as being due to a Doppler shift. The acoustic version of the Doppler shift is familiar to us as the change in pitch of passing cars. Cars coming towards us have a higher pitch than ones going way, since sound waves from cars travelling towards us are bunched up, or have shorter frequency. The same effect happens to light from a moving object. If a light source is moving towards us the light waves are bunched up and so bluer, and if it is moving away they are longer and so redder. Slipher suggested his observations meant that the nebulae were receding from us. Slipher even went so far as to suggest that this implied the Universe itself was expanding.

But Slipher did not know the distance to these nebulae. In fact the nature of the nebulae was still in dispute. Some thought they were clouds of gas floating around our galaxy, while others thought they may be galaxies in their own right. In the former case it could be argued that, although space may be infinite, the distribution of matter only extended as far as the galaxy and these clouds. This was the 'Island Universe' scenario.

But if these small nebulae were other galaxies like our own, filled with billions of stars and tens of thousands of light-years across, then they must be at huge distances away from us. And if we could see a few other galaxies, why should there not be more, an infinite number of them distributed throughout an infinite space?

Measuring the size of the Universe

It was Edwin Hubble who, in 1924, solved the problem of the nature of the nebulae and the size of space. Hubble was able to gauge the distance to the nebulae by showing that some contained variable stars, the 'Cepheid variables', whose variability in our own galaxy was tightly related to their brightness. Calibrating off the Cepheid variables in our galaxy, Hubble could estimate the intrinsic brightness of the Cepheid variables in one of the largest nebula, M31, and by comparing with their observed brightness estimate their distance. He found that this nebula was over a million light-years away. The Universe was suddenly a very big place.

In 1929 Hubble went further and began to compare his estimates of the distances to these galaxies to the recession velocities measured from the Doppler redshifts found by Slipher. Despite only having a few galaxies to work with and large uncertainties in the measurements,

Hubble claimed that his results showed a linear relationship between the distance to a galaxy and its recession velocity. Galaxies twice as far away from us were moving twice as fast away from us. In addition he found the motion of the galaxies in different directions was the same. Subsequent observations confirmed his discovery of this distance–velocity relation, now called Hubble's Law, and the constant of proportionality between distance and velocity was later called the Hubble parameter.

In fact Hubble's leap of faith that his data showed a linear relation between velocity and distance and that the Universe was expanding was not without precedent at the time. In 1915 Albert Einstein had finally unveiled his General Theory of Relativity, which would replace the Newtonian view of gravity with a new one based on the curvature of space and time.

Einstein had quickly appreciated that his new theory of gravity could be used to tackle the question of what the Universe looked like globally, which Newtonian gravity had so dismally failed to do. To solve the complex equations, Einstein assumed a solution with rotational and translational symmetries and, in 1917, found a model of the Universe which was spatially finite with no boundary – like the surface of a sphere. But Einstein's model was unstable. Gravity, being universally attractive, wanted to collapse the model universe. Einstein found that he could also make his universe expand, but both of these options seemed to him to be unsatisfactory. Slipher's discovery was not known widely and the Universe at that time appeared to be static. To make his model stable Einstein noticed that his equations allowed for an extra term he had previously neglected. This term permitted a universal repulsion which would counterbalance the attractive nature of gravity, and his model of the Universe could be made static. This extra term has subsequently been called Einstein's 'cosmological constant'. Unfortunately for Einstein, his model was still unstable. Any slight change in either the repulsive term or the attraction of gravity, by adding more matter or increasing the cosmological constant term, would cause the model to expand or contract again.

In 1917 the Dutch mathematician Willem de Sitter found another solution to Einstein's equations of gravity which again made use of the cosmological constant. In fact de Sitter's model had no matter in it at all and so would only expand. De Sitter further showed that there would be a linear proportionality between distance and velocity in his model. This was the first prediction of Hubble's law, but only for an empty universe. However, unlike Einstein, de Sitter maintained that relativity implied that the Universe must be dynamical and not static.

In 1922 the Russian physicist Alexander Friedmann, using the same symmetries as Einstein, found the general equation governing the evolution of a relativistic Universe and showed that it must be dynamic; expanding or contracting just as de Sitter had maintained.

When Hubble's observations, and Friedmann's prediction, become more widely known Einstein was understandably dismayed that he had not appreciated that the Universe itself could be dynamic and made the first prediction himself. He disowned his cosmological constant and famously dismissed the whole thing as 'my greatest blunder'. However, he did not mean that the introduction of the cosmological constant itself was a mistake, since it is consistent with Relativity; rather that he had missed an opportunity. Indeed, having drawn attention to the possibility of the existence of a cosmological constant the issue then became where was it? Clearly Friedmann's model looked more like the real world than either Einstein's or de Sitters and so this constant was either not there or was very small. But having introduced the cosmological constant it would prove hard to ignore it again. In fact it is arguable that far from being his greatest blunder, discovering the cosmological constant may have been one of Einstein's greatest achievements.

1.3 FOUNDATIONS OF THE BIG-BANG MODEL

In the intervening eighty years since Slipher's 1920 discovery of the expansion of the Universe, Friedmann's 1922 development of a dynamic model and Hubble's 1924 measurement of the distance to the galaxies and discovery of Hubble's law, cosmology has changed radically. Hubble had only a handful of local galaxies to measure distances and recession velocities. But modern galaxy surveys contain millions of galaxies and soon will have measured the redshifts of billions. All of these observations consistently show the same thing seen by Hubble, that distances are proportional to recession velocity. However, the value of the Hubble parameter was for many years a major issue in cosmology and observational measurements of distances could easily disagree by factors of two. Cosmology was for a long time data-starved.

With the advent of advanced technology, larger telescopes and electronic imaging rather than photographic plates, cosmology has turned from data-starved to data-rich. Observational estimates of the parameters of the cosmological model are now regularly measured at levels of a few per cent, and are found from a number of independent methods. While only 15 years ago there was a vast number of diverse models of the Universe (probably hundreds under discussion) and some of the basic principles of our understanding of the Universe were open to debate, in the intervening time this has all changed. In the following sections I will try and outline the development of these changes, from the introduction and establishment of the Big-Bang model of cosmology, its extension with 'cosmological inflation', and finally its current incarnation as the Standard Model of cosmology.

The modern Standard Cosmological Model is based on the older and highly successful Big-Bang model, developed from the 1930s onward. The main difference between the two is in how the initial conditions of the observed Universe are set. I return to this interesting issue in Section 1.5. Here we shall see why the older Big-Bang model became the accepted model to understand the general features and evolution of the Universe.

The emergence of the Big-Bang model marked the acceptance of relativistic models as the right way to describe the Universe. In science we always hope for a number of competing theories which can be compared with observations to decide which is correct (or at least closer to the observations). In the case of relativistic theories of the Universe the Big-Bang model competed for many years from the 1940s until the 1960s with the Steady-State Theory of Herman Bondi, Thomas Gold and Fred Hoyle. The Steady-State Theory was based on the idea of extending the spatial symmetries used by Einstein to solve his gravitation equations to include time. Not only would every point in space look the same, but every point in time should look the same. To achieve this they proposed a model which looked very much like the de Sitter model, but with constant matter density rather than with a cosmological constant. However, a constant density of matter in an expanding Universe would require its spontaneous creation to fill in the expanded volume in a way which was never quite explained. The Steady-State Universe was already in trouble in the 1960s when it was shown that quasar number densities changed over time, and finally killed off with the discovery of the cosmic microwave background (CMB), a thermal remnant from the hot Big Bang. Here I outline the three main observational pillars of the Big-Bang model: the expansion of the Universe, Big-Bang nucleosynthesis and the CMB.

(i) Expansion of the Universe

Slipher and Hubble's observations of the recession of the galaxies showed that they were moving away from us in the same way in all directions; there was an angular symmetry in the motion. Hubble also showed the recession increased with distance. Both factors can be accounted for if the distance between any two points in the Universe increases by an overall scale factor; a natural consequence of a relativistic model of the Universe. In addition, this expansion will look the same from all points, building in translational symmetry. This seems to imply preferred observers, who see the expansion the same in all directions, whereas General Relativity does not have preferred observers. The symmetry of the expansion arises because we have set the initial conditions for the model such that there are preferred states which see the symmetry. We will return to this later.

The expansion of the distance between points can also be a cause for concern. Is space itself being stretched? General Relativity only tells about the relation between points, not about the interval between them. Empty space is empty space so there is nothing to stretch.

Finally, if we extrapolate the expansion backwards in time the distance between any two points will tend to zero, and the density of points will become infinite. This has been interpreted by some as the actual creation of the Universe. In fact, as we shall see, our understanding of the physics of the early Universe breaks down long before this, so this is extrapolation beyond what we know.

(ii) Big-Bang nucleosynthesis

In the 1940s Russian physicist George Gamow, and later in 1957 Geoffrey and Margaret Burbidge, William Fowler and Fred Hoyle, developed the idea that if the Universe had been smaller in the past any radiation in it would have been at a correspondingly higher temperature. At some time in the past the Universe would have been hot enough to initiate thermonuclear fusion, just as it had recently been shown to power the Sun. But while stellar nucleosynthesis could explain the production of heavier elements, it had failed to explain why nearly all stars are made of around 25% helium. Gamow's calculations, and subsequent refinements, were able to show that given an initial abundance of hydrogen in the Universe, thermonuclear fusion in an expanding universe would spontaneously proceed to form deuterium, helium, lithium and beryllium in the first few minutes of the Universe. The relative abundances of these predictions, and in particular the 25% abundance of primordial helium, were compared with and found to be in very good agreement with the measured primordial abundances.

The relative abundance of the heavier elements depends rather sensitively on the initial density of baryons in the form of hydrogen. Taking the observed abundances of primordial elements implies a density of baryons that would contribute on around 5% of the value needed to make the Universe spatially flat. This startling discovery was the first indication that normal matter was only a small part of the Universe.

Knowing the temperature that primordial nucleosynthesis would take place at, and by estimating the density of matter required to produce enough Helium, Gamow was able to predict that the radiation at the time would have cooled by today to a few degrees Kelvin, leaving behind a microwave remnant of the early Big-Bang Universe.

(iii) Cosmic microwave background

One of the most powerful arguments for the Big-Bang model, and the one that killed off the Steady-State model, was Gamow's prediction that if the Universe was in a hot enough state to initiate nucleosynthesis in its distant past then matter and photons should have combined at high temperature to form a plasma. As the Universe cooled, well after nucleosynthesis, this plasma would have broken down allowing atoms to form and the photons to travel freely across the Universe. Assuming that nothing got in the way of the photons they would travel unhindered until they hit a detector on the Earth. A detector on the Earth it would see a uniform bath of radiation, now in the microwave range, coming from all directions.

The serendipitous discovery of this cosmic microwave background (CMB) radiation by Arno Penzias and Robert Wilson at Bell Laboratories in 1965, and its correct interpretation by Robert Dicke and Jim Peebles at Princeton, was seen as conclusive proof that the Universe had been hot and in thermal equilibrium, and that all that had happened to the radiation is that it has cooled due to the expansion of the Universe. The discovery by Penzias and Wilson of the CMB led to their award of the Nobel Prize for Physics and the establishment of the Big-Bang model.

1.4 THE INITIAL CONDITIONS OF THE UNIVERSE

Having observationally established the Big-Bang model of the Universe, attention naturally moved towards the events surrounding the very earliest moments of the model. But if one extrapolates the model backwards in time the scale factor goes to zero, implying that the density of the Universe everywhere becomes infinity. Some have mistakenly taken this to be an actual model for creation (hence the name Big Bang, a term of ridicule coined by Fred Hoyle) and assumed that one can ask no more. In fact the Big-Bang model does not explain the origin of the Universe: it is a model for its subsequent evolution. An analogy can be made with the theory of projectiles which can accurately describe how a cannon ball will travel through the air, but does not provide us with an explanation for how cannons work. To see why we cannot extrapolate the Big-Bang model back to the start we need to consider the limits of our knowledge of nature.

The quantum-gravity era

At high enough energies we expect all of the known laws of matter and spacetime to break down. In particular we expect that when energies become high enough, or on very small scales, gravity should come under the rule of quantum physics. Consider a massive particle. In General Relativity there is a length-scale, the Schwarzschild radius given by the mass of the particle, which tells us when we must consider the effects of curved space and time. This is the size the particle or object would be if it were a black hole. Usually the size of the object is much bigger than the Schwarzschild radius and we ignore space and time curvature. Quantum physics also tells us that the wave-like nature of this particle can be associated with a quantum wavelength, the de Broglie wavelength, after the French physicist Louis de Broglie. This is inversely proportional to the particle's mass (or energy). If we increase the mass (or energy) of the particle the Schwarzschild radius will increase while the de Broglie wavelength will decrease. At some point the de Broglie wavelength will become smaller than the Schwarzschild radius – we now have a quantum object where we need to consider the effects of curved space time. This happens at a scale called the Planck length where we expect quantum effects on spacetime itself to become important. Unfortunately we do not yet have a theory for how to combine quantum theory and General Relativity; a theory of 'quantum gravity'. This situation will be explained in more detail at the start of Shahn Majid's chapter in this volume.

An example of this happens at early enough times in the Universe when the energy-density of the CMB photons will increase to a point when classically they should have formed black holes. This is known as the 'quantum-gravity era'. Here the physics has become so extreme our knowledge of the properties of matter and gravity completely breaks down. As a result the Big-Bang model breaks down and we cannot use it to predict what happened before. Instead we must rely on more speculative theories to understand what may have happened. Our sole guiding principle in this regime is that nothing we speculate about happening here can conflict with what is already known about the Universe. There are a number of speculative research paths in search of quantum gravity. One of these is 'superstring theory', where all fundamental particles and spacetime are made of extended string-like objects. We would therefore like to know, for example, if superstring theory could yield a consistent picture of the early Universe.

Inflationary cosmology

By 1980 it was becoming clear that the initial conditions of the Big-Bang model of the Universe were a serious problem. As we noted earlier, when Einstein and later Friedmann began to look for models of the Universe they made some strong assumptions about the symmetry of the Universe to help solve the equations of General Relativity. Later on solutions were found that relaxed these assumptions and were less symmetric. But subsequent observations have shown that our Universe is highly symmetric. How did this happen?

In the early 1980s American particle physicist and cosmologist Alan Guth was looking for a solution to another problem. In the late 1970s and early 1980s speculative particle physics was predicting some exotic leftovers from the early Universe. If these theories were right, why were these remnants not around today? Guth's solution was to make the Universe rapidly expand for a short time to dilute these relics away. The reason we do not see them today is that they are very rare.

As well as the relic problem there had been another problem since the discovery of the CMB radiation. One of the symmetries of the Universe assumed by Einstein was isotropy: that the Universe looks the same to us in all directions. This assumption was beautifully confirmed by the CMB, which showed the temperature was the same to one part in a hundred thousand in every direction. But how did it get that way? Usually we assume the materials with the same temperature have been in thermal contact at some point. But according to the Big-Bang model points separated by more than one degree on the sky at the time the CMB was formed would never have been in causal contact and so could not have thermalised.

This causality problem arises because of a mismatch between the distance between points taking part in the expansion of the Universe and the distance light can travel since the initial singularity. In the Big-Bang model all of the mass and energy in the Universe is acting to slow down its expansion – reflecting the familiar experience that gravity is always attractive. If the expansion is slowing down, light emitted from one place will be able to catch up and overtake another point. This implies that any two points which are not yet in contact will never have been in the past.

To solve the causality problem we need to accelerate the expansion of the Universe. In this case if a photon is emitted from one point it may never catch up and overtake some distant point if the distance to the second point continues to increase faster than the distance light can travel. But if the expansion is accelerating when we go back to earlier times the rate of expansion must become arbitrarily slow. This implies that at earlier and earlier times any two points must at some time have been in causal contact, solving our causality problem. So how do we make the Universe accelerate?

The deceleration or acceleration of the expansion of the Universe is governed by the sum of energy-density and pressure in the Universe. In Newtonian gravity, the force of gravity only depends on the mass, and since mass is always positive we always get an attractive force. But in Einstein's theory of gravity the curvature of space and time, which gives rise to the force of gravity, responds to the total energy which is composed of both rest-mass energy and thermal (or pressure) energy. While the rest-mass energy of particles is positive, there is no fundamental reason why the pressure should be positive. A negative pressure can give rise to repulsion and hence acceleration of the expansion of the Universe.

But if normal matter has positive energy and pressure what type of substance violates this? In fact there is no physical reason why we cannot have negative pressure. We have already seen that Einstein's cosmological constant can lead to accelerated expansion in the de Sitter Universe. However we cannot use the cosmological constant here, since we only want accelerated expansion for a short time at the start of the Universe. It must switch off so that the Universe can enter a Big-Bang phase and agree with observations. We need a dynamical process that can be controlled. In fact we already know of one example that fits the bill.

In modern physics fundamental particles are the excitations of quantum fields. It is possible to suppress these particles, and their momentum, by suppressing the excitation of the fields. This has been demonstrated in the laboratory by the 'Casimir effect', named after the Dutch physicist Hendrik Casimir who first suggested the experiment in 1948. In the experiment, two metallic plates are set parallel to one another. The electromagnetic quantum fields between the plates have a discrete set of modes that can be excited, while those outside have effectively a continuum of excitable modes. This means that the pressure generated from particle momentum between the plates is less than that of those outside. The effect is a net quantum pressure pushing the plates together. Hence the Casimir effect, due to pressure of the quantum vacuum, allows negative pressure. This implies that to make inflation work we need to use quantum fields.

Having seen that cosmological acceleration can solve the causality problem let us see how it deals with the other problems of the

Big-Bang model. As we have noted, the first relativistic models of the Universe assumed high spatial symmetries: rotational symmetry around any one point, and translational symmetry between points. But the Universe could have been highly asymmetric. Indeed, locally we see large asymmetries due to the structure found in the distribution of the galaxies. Inflation can solve this problem as well, since the rapid expansion will also flatten out any non-uniformity in the distribution of matter and energy. After inflation we can expect the Universe to look very spatially symmetric.

Although Einstein's gravity allows space to be curved, it is not obvious to us that it actually is on a large scale. This is because the density of the Universe is close to that required to make it spatially flat. This is unexpected in the Big-Bang model, since any initial curvature of space in the early Universe should have grown as the Universe evolved. If we extrapolate back to the quantum-gravity era it seems the spatial curvature of our Universe was fantastically close to flat. How did this happen? Inflation solves this by increasing the size of the Universe by a fantastic amount. A good analogy here is with a balloon. Globally we see a balloon is round. But if we look at a small patch of the balloon as it inflates, locally it looks like it is getting flatter as the balloon expands. If we imagine the radius of the balloon becoming infinite, a patch on the surface would appear flat, just as the surface of the Earth looks flat because its radius is so large. Similarly we can expect that if the Universe has expanded by an enormous factor, the spatial curvature would have all but vanished. Hence a generic prediction of cosmological inflation is that on the large scale, the Universe should look spatially flat. As we shall see this prediction has unexpected consequences.

The standard Big-Bang model provides an accurate description of the expansion of the Universe, but not the origin for the expansion. In the inflationary model, the present observed expansion is the residual expansion due to the accelerating force of the inflationary era. In our cannonball analogy, inflation provides us with a mechanism for how the cannon works. When inflation ends, the Universe enters a

Big-Bang phase and the expansion continues, just as a cannonball's trajectory continues after it leaves the cannon.

While inflation has been amazingly successful at solving a wide range of problems associated with the initial conditions of the Big-Bang model, most of these could be regarded as being by design. During inflation the Universe is expanded by a large enough factor to solve the causality, symmetry and flatness problems. But these problems have been known for a long time. Maybe we've just been very clever in finding a single solution for them all. What is needed is a prediction for an observation we have not yet made. This would be a true test of the inflationary scenario. But before we find such a prediction, it is worth looking a little more closely at the flatness problem. At the time that inflation predicted the Universe should be spatially flat it was not accurately measured. Although inflationary models with some spatial curvature were proposed, most inflationary cosmologists would have said that spatial flatness was a strong prediction of inflation, due to the large factor inflation has to expand the Universe. If the Universe was not observed to be very close to flat I believe many cosmologists would begin to doubt inflationary cosmology. In Section 1.6 we'll look at what observations of the cosmic microwave background say about this.

The origin of structure in the Universe

Just after the first models of inflation were proposed, a remarkable consequence was discovered – inflation could provide a mechanism for the origin of structure in the Universe: the distribution of galaxies, galaxies themselves, stars and planets. In 1945 the Russian physicist Evgenii Lifshitz had shown that any initial variations in the Big-Bang model would grow under the pull of gravity. But although the Big-Bang model could account for the growth and evolution of structure, it gave no mechanism for where the initial structure came from.

Within a few years of Guth's original proposal of inflation, Guth and So-Young Pi, British cosmologist Stephen Hawking and American cosmologist Paul Steinhart had shown that the rapid expansion during inflation must vary slightly in different places due to quantum fluctuations of the field driving inflation. This would give rise to variations in the expansion rate of the Universe during inflation, leading to variations in the density of the Universe. The result was a prediction of how the relative amplitude of structure varies with scale, and how the amplitude of structure depends on the inflationary model.

This must rank as one of the most profound and significant insights in science – the implication of inflationary cosmology is that the largest structures in the Universe, with its complex array of galaxies, stars and planets, arose from physical processes on the smallest of scales. If we are looking for evidence of quantum processes we need only look to the skies.

As well as variations in the density of the Universe Russian cosmologist Alexie Starobinsky also realised that inflation would give rise to gravitational waves. Gravitational waves arise in Einstein gravity due to waves in the structure of space and time. Even though we do not yet have a theory of quantum gravity, it is still possible to approximately describe the quantum fluctuations of spacetime gravitational waves. Again the amplitude of these waves and how that changes with the wavelength of the wave can be related to the details of the inflationary model. These predictions, of the amplitude of variations in the density of the Universe on different scales and a relic gravitational wave background, provide us with new and powerful tests of inflation.

Outstanding problems with inflationary cosmology

Today inflation is still a paradigm and not a full-fledged theory. There is no fundamental physics underlying inflationary models, unlike say the Big-Bang model which is based on General Relativity. Inflation takes many of its ideas from 'quantum field theory', but the fields it uses are not directly related to the known fields of nature. If it turns out to be true inflation implies there is a wealth of physics beyond what we currently understand. One of the major aims of current research is to link inflation with what we know about particle physics.

Without a fundamental theory behind it, the question of how an inflationary period arose in the first place in the primeval Universe is open to speculation. To inflate even a small patch in the early Universe to look like the Universe we see around us today puts some restrictions on what came before inflation. For instance any initial patch must have been sufficiently uniform on average to allow inflation to start in the first place. How did this arise? Andre Linde at Stanford University in California has suggested a scenario called 'chaotic inflation', where the initial conditions of the Universe are random. In this model the Universe is much larger than we have imagined, and partitioned into domains with differing conditions. We live in a region where the right conditions for inflation have arisen by chance. This region then inflated and supplied the right conditions for life to form and ask the question 'where did we come from?'

A problem with Linde's proposal is that at this point the scientific process may start to break down. Since we cannot probe these other domains, unless they somehow impinge on our Universe, this model may not be testable. But it may be that inflation will fit into a more fundamental theory which explains how the right conditions could arise, and which can be tested. At present it is not clear if this is the case. However, the fact that inflation does such a neat job of giving a simple explanation for a diverse range of problems with the older Big-Bang model, and that some of its consequences could have been falsified already seems to most cosmologists like reasonable evidence that something like inflation did happen. What is needed is a clear smoking gun.

1.5 THE STANDARD COSMOLOGICAL MODEL

During the several decades from the 1920s to the 1970s, the Big-Bang cosmological model was developed and established as an accurate description of the Universe both today and over most of its history. In the 1980s inflationary cosmology appeared to explain the initial conditions for the Big-Bang model, but, although compelling, still needs firm observational evidence and a firmer theoretical basis. Around the turn of the millennium the level of accuracy and the consistency between diverse and independent observations, combined with a consistent theoretical model tying all of these observations together, laid the basis for what I will call the Standard Model of cosmology, in the same way that in the 1970s there emerged a Standard Model of particle physics.

Just what is meant by a 'Standard Model' in physics is a little hard to say. Here I will mean that it is a model agreed upon by the majority of cosmologists, against which any competing model should be compared, and against which all observation should be checked for consistency. Significantly, it is a model with which we can make predictions to be tested by future observations. This does not mean that some day the model we are using will not be overturned or superseded by a better model. Only that it seems the model has real explanatory and predictive power which should be exploited. But despite this confident mood that real progress has been made in cosmology, these developments have been double-edged. The Standard Model of cosmology has turned out to be a very strange one, in fact the model that no one wanted. To explain why, we need to look in detail at the ingredients of the model.

The ingredients of the Cosmological Model

Since the Standard Cosmological Model builds on the success of the Big-Bang model, its ingredients start with Einstein's theory of gravity, General Relativity, which forms the backdrop of a dynamical Universe filled with energy and matter and which governs its expansion and contraction. In addition we also include Einstein's assumption of spatial symmetries; rotational symmetry about a point and translational invariance between points.

The second ingredient is the Standard Model of particle physics. Even though we don't yet have a theory which unifies gravity and particle physics it is a non-trivial observation that a relativistic model of the Universe and particle physics are consistent (although later we will find a significant counter-example). It would not be unimaginable for particle physicists to find a particle whose mass-density was observed to be so high it should have caused the structure of spacetime to have folded back in on itself and re-collapsed the Universe long ago. But so far no such particle has been seen. In fact, as we have seen in Section 1.3, setting nuclear physics in an expanding Universe yields Big-Bang nucleosynthesis and we find that the contribution of baryonic matter can only account for around 5% of the total contents of the Universe.

The next ingredient, making up some 20% of the Universe, is an exotic form of dark matter which is not currently part of particle physics. The idea of dark matter has been around since 1932, when the Hungarian-born astronomer Fritz Zwicky, working at the Californian Institute for Technology, found that galaxies in nearby clusters of galaxies were moving too fast to be bound by their gravitational mass and should have flown apart long ago. He proposed that there was more mass than could be seen – now called dark matter. In 1932 Einstein and de Sitter also authored a joint paper where they set the cosmological constant to zero. Since the observed abundance of stars in the Universe was too small compared with their model, they argued that most of it was in the form of Zwicky's dark matter. In the 1970s and 1980s the Russian School of Physics, led by Yakov Zel'dovich, pursued the idea that the dark matter could be massive neutrinos, particles which take part in nuclear decay. But by the early 1980s it had become clear that these fast-moving particles would not be gravitationally bound into galaxies and so would have prevented their formation. Around this time Jim Peebles at Princeton looked at new ideas in particle physics and proposed the dark matter was not 'hot', relativistic when it formed, but cold, or slow-moving when it formed. A new proposal in particle physics, that every force particle had a matter counter-part and vice versa, called supersymmetry, provided a favourite candidate dark-matter particle, the neutralino.

More disturbing than the introduction of dark matter is the next ingredient: a mysterious force which looks like Einstein's cosmological constant and is now called dark energy. This seems to make up a

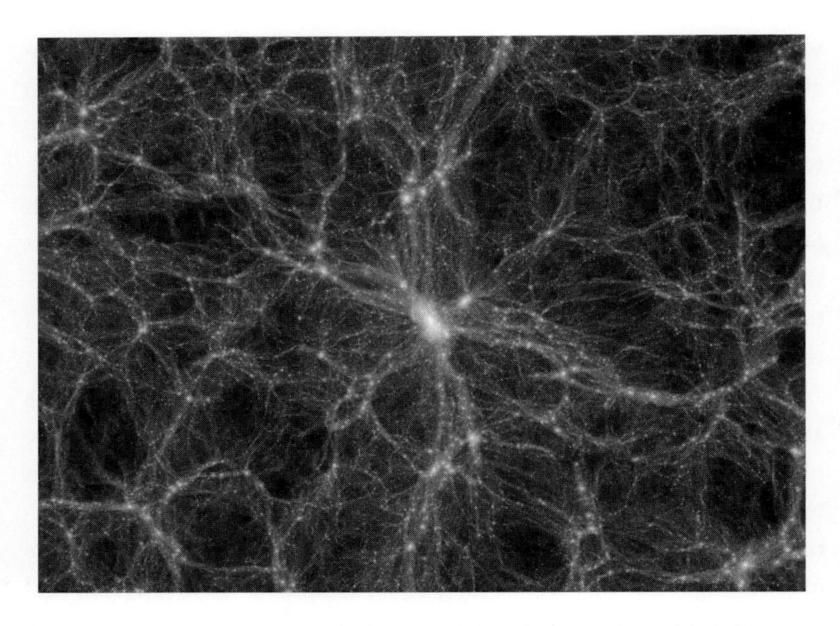

FIG. 1.1 Simulation of the spatial distribution of a cold dark-matter Universe. The simulation is centred on a dark-matter cluster which is connected to other clusters by filaments. Along the filaments are beads of smaller dark-matter haloes, while between the filaments are planar sheets of dark matter. Image courtesy of the Virgo Consortium (www.virgo.dur.ac.uk).

full 75% of the energy budget of the Universe. We shall return to the observational motivation for this ingredient later.

Finally, as well as the background spacetime and material ingredients of the Cosmological Model, we need a set of initial conditions at the start of the evolution of the Universe. The current best idea for setting these initial conditions is the cosmological inflation hypothesis.

If the Standard Model of cosmology holds true, the Universe is not as we see it – visually dominated by starlight tightly bundled into galaxies and spread across space. Instead, the Universe actually looks like the simulation shown in Figure 1.1, dominated by cold dark matter and whose evolution is governed by the dark energy. The galaxies we see lie within smaller dark-matter haloes embedded within dark-matter clusters and strung along filaments. One of the current aims of cosmology is to actually detect this complex structure and understand its relation to galaxies.

1.6 THE EVIDENCE FOR THE STANDARD COSMOLOGICAL MODEL

Just as there are three main pillars of the Big-Bang model of the Universe, there are three main observational pillars establishing the Standard Model of cosmology: these are the evidence from Galaxy Redshift Surveys (GRSs), the cosmic microwave background (CMB) radiation and the light from Type Ia supernovae (SN1a).

(i) Galaxy redshift surveys

Hubble's law says that the distance to a galaxy is proportional to its redshift. Hence we can use the observed redshift of a galaxy to estimate its distance. Since we also know that galaxy's position on the sky we can plot its position in three dimensions and make a map of the galaxy distribution across the Universe. Figure 1.2 shows the spatial distribution of over a quarter of a million galaxies in the Anglo-Australian 2 degree Field Galaxy Redshift Survey (2dFGRS). The survey maps out two patches on the sky which, in three dimensions, appear as two cones centred about us. Immediately we can see two things. Firstly that the distribution of galaxies is not uniform, there is a clustering pattern. Secondly this pattern looks similar on both sides of the surveyed patches, indicating that on large scales there is statistical uniformity. This would seem to indicate that Einstein's assumptions about the symmetries of the Universe, in a statistical sense, are true.

In detail the pattern sketched out by the distribution of galaxies is highly complex. Galaxies are not distributed uniformly, nor are they distributed randomly. Instead they trace out groups and clusters of galaxies made of tens to thousands of members. Spanning these clusters are filaments, or strings, of galaxies and between the filaments there are walls, or sheets, of galaxies. Explaining the origin of this pattern has proven surprisingly difficult. However the cold

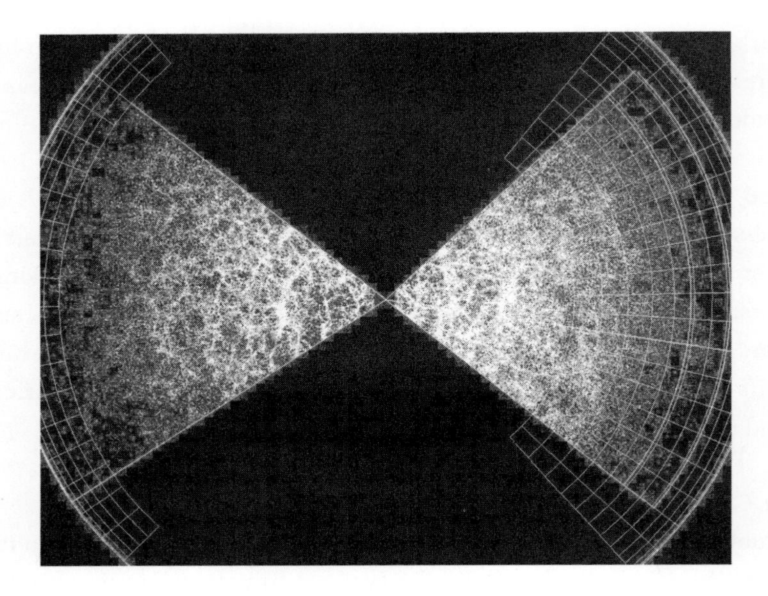

FIG. 1.2 The distribution of galaxies in the 2 degree Field Galaxy Redshift Survey (2dFGRS; www2.aao.gov.au/ TDFgg/). Each slice is a different direction on the sky centred on our Galaxy. The survey is around four billion light years across.

dark-matter scenario seems to naturally give rise to this sort of pattern of clusters, filaments and sheets and it's not hard to believe that, at least on large scales where there has not been time for the dark matter and galaxies to segregate, the galaxies trace the underlying dark-matter pattern.

If the Universe is mostly filled with dark matter and there is also radiation around from the CMB we can make a simple but significant prediction. While matter seems to dominate the Universe today (there is more energy in the form of matter than radiation), in the past the reverse must have been true. This is because while the density of matter and photons is decreased by the expansion of the Universe, the radiation also suffers from redshifting. This change in colour is a change in the energy carried by the radiation so the total energy in radiation decreases faster than that of matter. If we go back in time, the energy in radiation must grow faster than that of matter and at some point exceed it. At this time we are now in a radiation-dominated Universe.

Let us consider the evolution of a clump of dark matter in the 'radiation era'. Specifically, consider an overdensity of dark matter and radiation whose size is much larger than the distance light can travel. Radiation has not had time to escape from this overdensity, which will then evolve like an isolated universe governed by the local density. The overdense region will not expand as fast as the rest of the Universe and so the overdensity will grow. Now consider a cluster of dark matter whose size is less than the distance light can have travelled since the initial singularity. In the radiation era, the gravitating mass-energy in the cluster is negligible compared to the energy in the radiation. Since the radiation does not interact with dark matter it is able to cross the cluster and escape, and so there is negligible gravitational force in the cluster to make it collapse under its own weight; the cluster does not grow. So on scales larger than the light-crossing time structure will grow in the radiation era, while on scales smaller than this it does not.

In the matter-dominated era the same argument still holds on large scales where distances are further than light can travel and the cluster grows. But on small scales in the matter-dominated era the majority of the mass-energy is in dark matter, not radiation, and so the cluster will collapse under its own weight. In the 'matter era' structure grows on both small and large scales. As there are no preferred scales in gravity, or for cold dark matter, structure grows at the same rate on all scales.

The significance for the pattern in the galaxy redshift surveys is that if we assume a simple initial distribution of dark-matter structures laid down in the early Universe – a continuous range of structure from low-density structures on large scales to high-density haloes on small scales predicted by inflation – this pattern is subsequently changed in the radiation-dominated era as structures on very large scales grow while those on small scales do not. A break is created in the range of structure at the scale over which light can have travelled

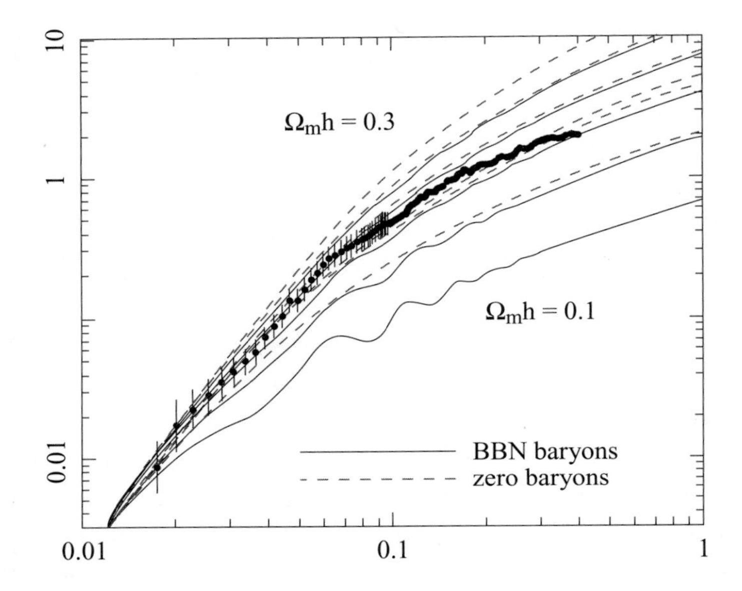

FIG. 1.3 The amplitude of structure as a function of scale in the Universe from the 2dFGRS. The upper axis is the fractional overdensity of the distribution of galaxies, which are assumed to trace the fractional overdensity of all matter. The horizontal axis is the Fourier wavenumber, k, which corresponds to the inverse scale of the overdensity decomposed into waves. Points are the 2dFGRS data with error bars, dashed lines are models with only cold dark matter and solid lines are models with cold dark matter and baryons. Image credit: J. Peacock.

at the time of the transition from radiation to matter domination. The time of the transition depends on the density of matter in the Universe so a measurement of this scale is also a measurement of the density of matter in the Universe. Since structure grows at the same rate on large scales this feature is frozen into the matter distribution and hence the distribution of galaxies today. Figure 1.3 shows the amplitude of structure as a function of scale. On large scales there is a continuous range of amplitudes from the initial conditions in the Universe. But on smaller scales there is a break and turn-over due to the suppression of growth in the radiation era.

Figure 1.3 also shows a series of models. The smooth curves are for models containing only dark matter and radiation. Already these models look very much like the pattern in the galaxy data. The break in the shape of the galaxy power spectrum can be compared with models of the dark-matter power spectrum and indeed there is a break at around the scales expected, and the overall shape is as expected. This is great news for cold dark matter, since it seems to confirm we are on the right lines. It is worth repeating here that reproducing the shape of the galaxy power spectrum is not easy – many alternatives to the cold dark-matter picture have been suggested, but most fail to reproduce the shape of the galaxy power spectrum.

By varying the amount of dark matter in the models, and so changing the position of the break in the power spectrum, we can measure the density of matter in the Universe. A convenient measure is how much matter is required to make the Universe spatially flat. From the break in the galaxy power spectrum we find there is only 30% of the matter density required to flatten the Universe. This is already six times larger than the maximum density of baryonic matter, predicted by Big-Bang nucleosynthesis, which could be in the Universe. Hence the majority of matter in the Universe must be non-baryonic dark matter.

But we know that the Universe also contains baryons today in the form of stars and gas. If we add hot baryons to the photons to form a plasma in the early Universe we see a change in the model predictions. Oscillations appear – so-called baryonic acoustic oscillations (BAOs) or 'baryon wiggles'. The origin of these wiggles can be traced back to evolution of dark-matter haloes embedded in the plasma of baryons and radiation. The gravitational attraction of the dark-matter halo will pull the plasma into it. But the pressure of the plasma will cause it to bounce out of the dark-matter halo. On the largest scales there will only be time for this to happen once, before the Universe changes from plasma to free photons and atoms. After this the photons will fly free and spread uniformly across the Universe.

However, the baryons will have formed a small shell around the dark-matter halo. The gravitational pull of the baryon shell will pull dark matter into it, producing a dark-matter shell. Eventually, over time, the baryons and dark matter trace out the same distribution of a halo with a small shell surrounding it. The baryons eventually collapse to form stars and galaxies. The distribution of these galaxies will follow that of the baryons and dark matter on large scales, and so around galaxy clusters we can expect to find a small (about 1%) excess of galaxies. This excess has recently been detected, at the predicted scale of 450 million light years across in the distribution of galaxies in the 2dFGRS and the SDSS. The shell appears as a small excess in the correlation of galaxies at this scale. The power spectrum of overdensities shown in Figure 1.3 is the Fourier transform of this, where such an excess will appear as oscillations, giving rise to the baryon wiggles.

Again this is an amazingly accurate verification of the Standard Model of cosmology, and shows that even the fine details of the model are correct. Just as exciting, the size of the baryon wiggles can be used to measure the density of baryons in the Universe. Observations of galaxy redshift surveys show that this is around 5% of the density required to flatten the Universe. This is remarkably close to the value predicted by Big-Bang nucleosynthesis, strongly suggesting we are on the right lines.

(ii) Cosmic microwave background

Along with the distribution of galaxies, a major source of information in cosmology comes from the cosmic microwave background. As the Universe contains structure, this should be reflected in variations in temperature of the CMB, as first pointed out by Rainer Sachs and Arthur Wolfe in 1967. These temperature variations were first discovered by the COsmic Background Explorer (COBE) in 1992, and led to its two lead scientists, Charles Bennett and George Smoot, being awarded the second Noble Prize for Physics for the CMB in 2007.

Although the pattern of temperature fluctuations in the CMB looks random there is actually a pattern which can be interpreted to give us information about the events when it formed. On large scales, greater than the distance light can have travelled between the formation of the Universe and the formation of the CMB, the only relevant force can be gravity. When photons break free of the baryons, some photons will be at the centre of dark-matter haloes. Here the gravitational field of the halo is at its greatest. According to General Relativity time runs slower in a gravitational field and so the wavelength of light is longer. In addition, to climb out of the dark-matter halo the photon loses energy and undergoes a gravitational redshift, again increasing the wavelength of light. In 1967 Rainer Sachs and Arthur Wolfe showed that the temperature change of light due to this was proportional to the gravitational strength of the dark matter. Hence on scales greater than one degree, when larger regions at the time the CMB was formed were not in causal contact, we are directly seeing the distribution of the dark matter reflected in the temperature pattern.

As we move to smaller scales, the distance between points on the CMB becomes less than the distance light can have travelled. Now causal physical processes can become important. In particular the interplay between the evolution of baryons, photons and dark matter becomes important. Just as we described for the creation of baryon wiggles on the largest scales, the baryons, as they become free of the photons, fall into the dark-matter haloes. On smaller scales there is enough time for the baryons to be blown back out of the halo by the build-up of pressure and then to fall back in again. As the baryons fall back in, the pressure builds up again until they are forced back out again. Baryons bounce in and out of the dark-matter haloes, under the oscillating forces of gravity and pressure. As the baryons fall in and the pressure builds up, the density of the baryons increases. The baryons are still weakly bound to the photons so the photon energy-density also increases and the temperature goes up. Similarly as the baryons bounce out, their density drops and the photon temperature falls. The bouncing of the baryons results in an increase and decrease in temperature which should be reflected in the CMB temperature fluctuations.

During this period the scattering which binds the baryons and photons together starts to break down and the photons start to seep past the baryons. High-temperature regions in the CMB start to leak

FIG. 1.4 The pattern of temperature fluctuations in the cosmic microwave background projected onto the celestial sphere seen by WMAP. The pattern of temperature fluctuations on scales greater than a degree directly traces the gravitational potential of dark matter. Those on smaller scales reflect sound waves in the plasma filling the Universe at the time. Image credit: M. Tegmark.

into low-temperature regions, blurring the distinction between them. On small angular scales this leakage smoothes out the pattern of the temperature fluctuations first noted by Joseph Silk.

This picture of the interplay between baryons, photons and the dark matter was worked out in detail in the 1970s and 1980s, most notably by Jim Peebles at the Institute for Advanced Studies in Princeton. By the 1990s the theoretical picture of the CMB temperature fluctuations was in an advanced state when the large-scale fluctuations predicted by Sachs and Wolfe were first detected by the COBE satellite.

The power of the cosmic microwave background, combined with analysis of the pattern of the galaxy distribution became quickly apparent when it became clear that most of the proposed theories

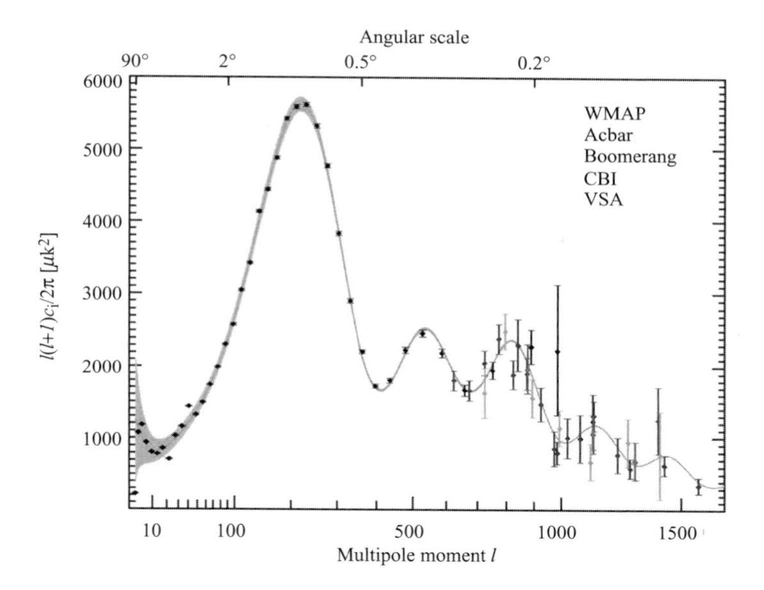

FIG. 1.5 The amplitude of temperature fluctuations as a function of angular wavenumber (inverse angular scale) on the sky measured by WMAP. The first peak at $l = 200$ is due to the first collapse of the photon–baryon fluid into dark-matter haloes. This happens on the scale that sound can travel in the plasma at this time. Given the known temperature and composition of the plasma this is a known physical distance. Source: map.gsfc.nasa.gov.

of dark matter, or alternative models for how structure arose in the Universe, were ruled out in the early 1990s.

This rapid progress in cosmology continued with a series of ground-based CMB experiments which helped pin down the first of the acoustic oscillations in the CMB temperature spectrum. But the most spectacular results came from the Wilkinson Microwave Anisotropy Probe, or WMAP. This was a relatively small satellite sent out to one of the Lagrange points – a point of stability away from the Earth and away from the Sun. Despite its size, away from the Earth and its atmosphere WMAP could steadily survey the whole sky. When the first year of the survey was complete and the temperature power spectrum measured, the agreement between theory and observation was astonishing. Figure 1.5 shows the results from the first year of data from WMAP and other data, compared with the best-fitting Standard Model of cosmology.

As well as measuring the total density of matter and the relative fraction in the form of baryons and dark matter, the CMB temperature pattern can also be used as a standard ruler to measure the geometry of the Universe, the shape of space. The first peak in the CMB temperature fluctuation power spectrum corresponds to red hot-spots in the CMB maps. These have a characteristic size reflected in the position of the first peak of about one degree. The reason for this characteristic size is that these hot-spots show where the plasma in the early Universe has just collapsed into the dark-matter haloes and heated up with an increase in pressure. On scales larger than one degree information carried by sound waves has not had time to travel through the plasma. As a result the plasma and dark matter can only feel the force of gravity and are still collapsing under gravity's influence. On scales smaller than two degrees the plasma oscillates in and out of the dark-matter haloes under the influence of gravity and pressure, as we described for the baryon wiggles.

Since the physical size of the largest hot-spots is set by the sound speed in the plasma which only depends on the temperature and composition of the plasma, we have a 'standard ruler' for the CMB. We also know the distance to the CMB, which depends on the composition of the Universe. With this we can form a triangle in the Universe, with ourselves at one point, and the other two points on either side of a CMB hot-spot. The apparent angle that the hot-spot subtends on the sky will depend only on the geometry of space (see Figure 1.6). If space is flat, the apparent angular size of the spots will appear as expected since the internal angles of the triangle will add up to 180 degrees. But if we live in a Universe with positive curvature, like a sphere, the apparent size will be larger than expected. Similarly if we live in a Universe with negatively curved space the spots will appear smaller than expected.

This way of measuring the geometry of space was first suggested by the astronomer and mathematician Johann Carl Friedrich Gauss.

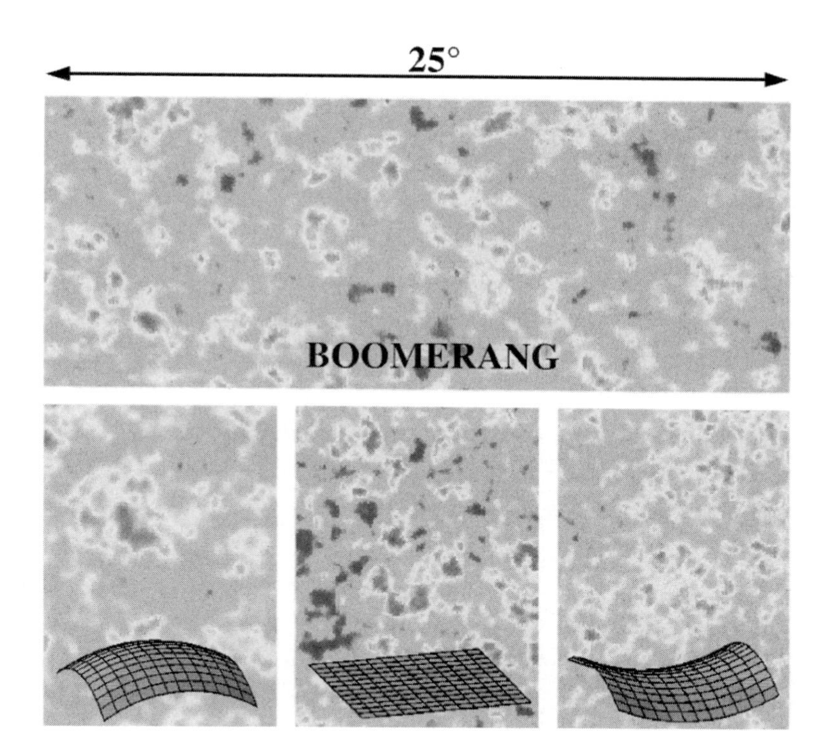

FIG. 1.6 Measuring the geometry of the Universe from the CMB. LHS: In a closed universe light beams converge and so the internal angles in a triangle add up to greater than 180 degrees. Middle: In a spatially flat (Euclidean) universe light follows straight lines and the internal angles in a triangle add up to 180 degrees. RHS: In a spatially open universe light paths diverge and the internal angles of a triangle add up to less than 180 degrees. For a fixed physical size at the CMB the angular size on the sky is bigger, or smaller for a closed or open universe. Source: www.astro.caltech.edu/~lgg/boomerang_front.

Gauss actually tried to carry out his idea to measure the geometry of space around the Earth by joining a land survey, so that he could measure triangulation points and see if their internal angles added up to 180 degrees. Of course we now know that the gravitational field around the Earth is too small to noticeably curve space. But the CMB shows his idea works.

In practice we compare the CMB temperature power spectrum against models with variable geometry, but the result of a spatially flat

Universe is that the first peak should be at one degree. From Figure 1.5 we can just read off the result at the first peak and we can see the CMB indicates the Universe is indeed flat. Given the result we found earlier, that the observed matter-density of the Universe is only 30% of that required to make the Universe flat, this leaves us with a new 'missing energy' problem.

(iii) History of the expansion from supernovae

As well as using a 'standard ruler' to measure the properties of the Universe, we can also use what we call 'standard candles'. This is a set of astronomical objects which have the same intrinsic brightness – just as we can use the apparent brightness of a light bulb with known wattage to estimate its distance, brighter if it is nearer and fainter if it is further away. The only problem is to then find such a 'standard candle'.

A good contender for a 'standard candle' is a subset of supernovae called Type Ia. Supernovae are the nuclear explosion of stars at the end of their lives. Type II supernovae are caused by the destruction of massive stars which have exhausted their hydrogen supply for nuclear reactions and left the 'main sequence' where stars spend most of their lives to become 'red giants'. A star's life as a red giant is limited. It is surviving on a small amount of helium which it is nuclear burning to make heavier elements. When this supply of helium is exhausted the nuclear reactions which supply the energy to heat up the atmosphere of the star and prevent its collapse come to an end. Without the thermal pressure to hold up the atmosphere it collapses under its own gravity onto the core of the star. The core itself collapses but then reaches a density where quantum effects prevent it from collapsing further. The core bounces and sends out a blast wave of energy which tears off the atmosphere, resulting in the supernova burst. However the strength of this burst depends on the initial size of the star, and so is not a 'standard candle'.

Type I supernovae arise in binary star systems, which make up about half of all stellar systems. Usually one star is more massive than the other. The more massive star will live life faster than its smaller companion and will become a red giant. If it is too small to become a supernova Type II, it will exhaust its nuclear fuel and then just die, blowing off its atmosphere when the core collapses to form a 'planetary ring'. The resulting core will cool to become a 'white dwarf'.

Meanwhile the less massive companion star will continue with its life, eventually exhausting its hydrogen fuel and becoming a helium-burning red giant. As the core contracts and heats up to reach helium-burning levels, the atmosphere expands, becoming red. As the atmosphere of the red giant expands, and the white dwarf is close enough, it can reach a point where it comes under the influence of the companion white dwarf's gravitational pull. The atmosphere starts to spill out of the red giant and flows towards the white dwarf. This material from the red giant will spiral down and collapse onto the white dwarf, see Figure 1.7. Slowly the mass of the white dwarf will increase as more and more material falls onto it. This cannot continue indefinitely. Eventually the mass of the white dwarf is so high the internal pressure has increased to a point where the material in the dwarf will undergo nuclear detonation, and the white dwarf will explode as a Type I supernova. Since the mass required to start the nuclear detonation is fixed, all Type I supernovae should have the same brightness.

In detail this is not quite true. The Type I come in two varieties, Ia and Ib. The Ib do vary, but the Ia do appear to be uniform. The Type Ia do have a slight variation in brightness, but this seems to correlate with peak brightness. Hence by measuring the peak brightness of the supernova a correction can be made to make a very uniform sample.

To use the Type Ia supernovae to measure the properties of the expansion of the Universe we can plot a supernova 'Hubble diagram' – that is we plot the brightness of the supernova which is a measure of its distance, against redshift which is a measure of how much the Universe has expanded since the light from the Type Ia was emitted. The relationship between the distance that light has travelled across

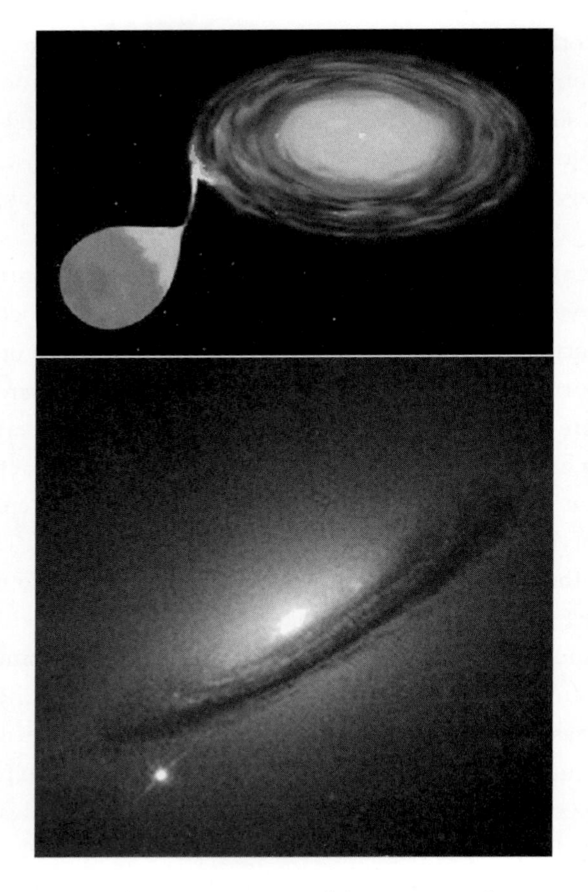

FIG. 1.7 Top: Artist's impression of the accretion of matter onto a white dwarf in a disc, fed by the matter from a companion red giant. Bottom: A Hubble Space Telescope picture of light from a supernova on the outskirts of a galaxy. Source: hubblesite.org.

the Universe and how much the Universe has expanded (measured by how much the light has been redshifted) depends on the cosmological model. If the Universe was static the distance would increase, but the redshift would be zero. In a matter-dominated Universe we have seen that the expansion of the Universe can only decelerate. If we trace the expansion backwards in the model we eventually reach a point where the expansion started – the so-called 'Big-Bang' singularity. Since this

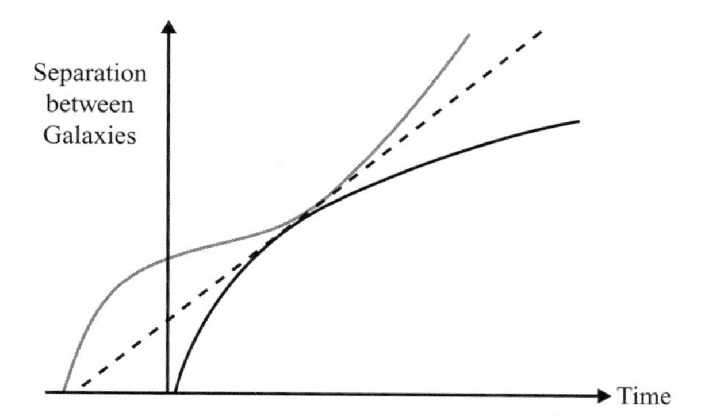

FIG. 1.8 The expansion of the Universe with time for different cosmological models. In a matter-only model the Universe can only decelerate, and the initial singularity is a finite time in the past (black line). For some models (for example a model whose energy is only in the curvature of space) the expansion is linear (medium line). For the same expansion rate today (measured by the Hubble rate) this model will be older than a matter-only model. If the model is accelerating today, the initial singularity is pushed back even further and the model is even older (light line).

singularity is only a finite time in the past, light can only have been travelling for a finite time, and as light travels at a finite speed, the distance it can have travelled is finite. If we plot the distance light can have travelled against the amount the Universe can have expanded (the redshift), as the redshift increases (and we are looking back in time) the distance travelled must become a constant. Beyond this constant distance light just has not had time to travel to us.

But if the Universe is in a state of constant acceleration and we keep the current expansion rate fixed, we can push the start of the Universe further back. This gives light more time to travel across the Universe for a given amount it can have expanded. In principle the Universe can be infinitely old in a model like this, and so the distance light can travel can increase indefinitely. Comparing these two models, for a given redshift (or factor the Universe has expanded) light will have travelled further in the case of an accelerating Universe compared to the case of the decelerating model. Figure 1.8 shows the different evolution of different models.

Now, let us return to the observations of Type Ia supernovae, where we are measuring the apparent brightness of the supernovae against their recession redshift. The result of these observations, first made be teams led by Saul Perlmutter at the Lawrence Berkeley Laboratory at the University of California and Adam Reiss at Johns Hopkins University in Baltimore, is that the Type Ia are fainter than expected for their redshift. We can interpret this as saying that they are fainter because they are further away than expected for a matter-only universe. In other words the supernovae results show the Universe must be accelerating.

1.7 THE DARK SIDE OF THE UNIVERSE

Combining our results from the CMB, galaxy redshift surveys and the supernovae we reach the conclusion that (a) the Universe is made up of 30% matter, (b) only 4–5% of it is normal (baryonic) matter, (c) the Universe is spatially flat so there is a missing 70% component which is not matter, (d) the expansion of the Universe is accelerating so there must be a substantial negative-pressure component. Figure 1.9 summarises these results.

How reliable are these results? Maybe we are mistaken to rely on these probes of cosmology and so our interpretation is incorrect. A few years ago this would have been a reasonable position. The results were new, and of course could have changed with the arrival of more and better data. Instead, in the intervening few years, these results, and hence their conclusions have only strengthened. In fact, one of the results was in an independent verification of a much older result. For many years astronomers have studied the properties of clusters of galaxies, made up from tens to hundreds, even thousands, of galaxies. These seem to be the largest stable structures of the Universe, and so should be representative of the constituents of the Universe as a whole. But every indication from clusters was that the total density of matter in the Universe is only 30% of that required for a flat Universe. The results from galaxy redshift surveys only gave this independent support.

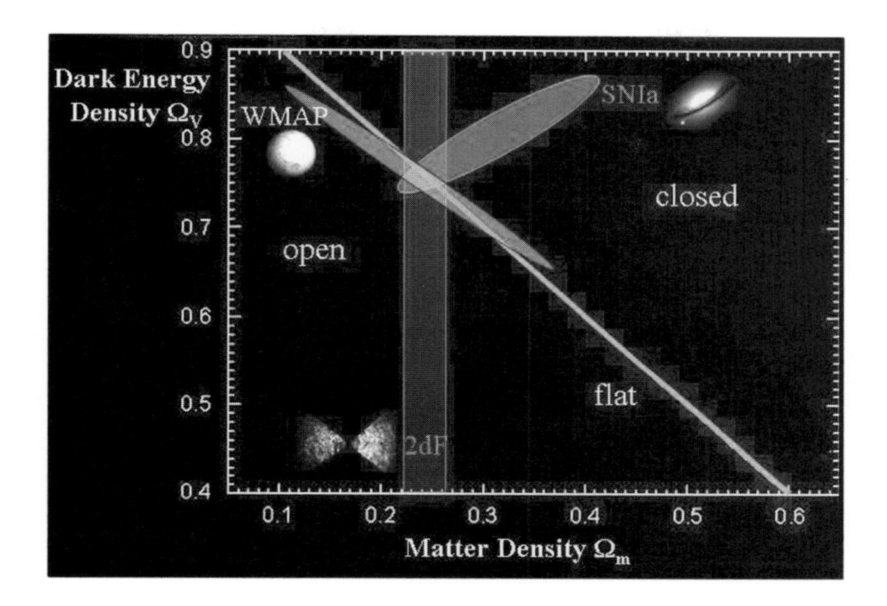

FIG. 1.9 The parameter space of dark energy density and matter density with constraints from observations of the CMB with WMAP, the galaxy distribution from the 2dFGRS and HST observations of supernovae Type Ia.

Further evidence was also from a long-standing problem with the most distant (highest redshift) galaxies. When astronomers studied the ages of these galaxies from their starlight, they found that the ages of these galaxies were greater than the age of the Universe in a matter-only model. One way out of this was to appeal to cosmic acceleration to make the Universe older than it looked (recall Figure 1.8 shows that the time between an initial singularity and the present with a fixed expansion rate is greater in an accelerating model). In fact the impressive thing about this model of cosmic acceleration is that all our observations appear to confirm it.

Before we proceed we must deal with the question of what the dark energy actually is. Later we shall address this in more detail and outline some of the leading proposals. But even before the data began to suggest a major negative-pressure component to the Universe, there was already a theoretical prediction for it.

As discussed earlier, Einstein had introduced his cosmological constant term to create a static (but as it turned out unstable) model universe. On learning that the Universe was actually expanding Einstein described the introduction of this term as 'my greatest blunder'. Presumably he meant he had missed the opportunity to predict the expansion of the Universe – surely the most dramatic manifestation of a dynamic spacetime. In fact Einstein had done nothing wrong in introducing this term, and as we saw de Sitter made use of it to make the first expanding model Universe. De Sitter's model was however empty of matter, but was accelerating. The reason for this was the constant energy-density of the cosmological constant. As the Universe expands we expect the energy-density to drop. If energy is conserved and the volume is getting bigger this makes sense. So how does the energy-density of the cosmological constant stay the same as the Universe expands? It does so because it has negative pressure. The negative pressure gives rise to a negative energy, which compensates for the loss of energy as the volume increases. Alternatively, we could start with the definition that the cosmological constant does not lose energy, and show that this implies it has a negative pressure. Either way, the cosmological constant and de Sitter's model of the Universe seems to be exactly what the data implies.

These results are robust. We have mentioned here three of the primary observations which have led us to dark energy, but we could drop any one of them and reach the same conclusion. That all three agree about dark energy should already give us some confidence. In fact there are many other observations, all of which lead us to the same conclusion. For example, we have already mentioned the ages of galaxies which seemed too old compared to a matter-only universe. In addition to the results from galaxy redshift surveys about the low density of matter in the Universe, the study of galaxy clusters had long suggested that the density of matter was low.

Cold dark matter

So what is the dark matter? In short, we don't yet know. As we've seen, observations suggest that the dominant form of dark matter is not baryonic, since the limits imposed by Big-Bang nucleosynthesis say that at most only 5% of the energy-density of the Universe can be in the form of atoms interacting in a known way. Observations also suggest that it is particle in nature, rather than made up of, say, black holes. One possibility is that the dark matter is neutrinos, tiny particles which are produced in weak-force interactions, such as nuclear decay. However cosmological observations had shown that their mass must be small to avoid making the density of the Universe higher than the observed value. In addition, since neutrinos only interact by the weak force they can fly out of gravitationally bound clumps, like galaxies, stopping the formation of the galaxies. This simple observation means that neutrinos were not the dominant form of dark matter. However, even if they are not dominant recent observations in Japan and the US have shown that the mass of neutrinos is not zero. In fact their mass-density could equal that of the baryons and contribute around 5% to that of the Universe.

The nature of the dark matter seems to be most likely a new particle. Conveniently, particle physics provides us with a leading contender with the right properties for dark matter. To unify the particles that make up matter and the carrier particles of force, particle physics introduced the idea of 'supersymmetry' which allows one type of particle to turn into another. But this idea needs the introduction of a partner particle for each matter and force particle. For the photon there is an as yet unseen photino partner. For the neutrino there is the force-carrier sneutrino. The lightest of these supersymmetric particles is actually a composite of the photino, the Higgsino and the zino, called the neutralino.

While the existence of these exotic particles may seem far-fetched, their detection, if they exist, may not be far away. The Large Hadron Collider (LHC) at Europe's CERN laboratory has enough power to reproduce conditions some 10^{-15} seconds after the Universe was born. At these high energies we can hope to see evidence for supersymmetry. If we do we can expect to produce supersymmetric particles, the lightest of which will be the neutralino. If we do, then we may just have produced our own dark matter.

Dark energy

The evidence of the flat geometry of the Universe measured by the CMB, low total density of luminous and dark matter, and the recent accelerated expansion history of the Universe all point towards a new component of the Universe which dominates the energy-density, but does not behave like normal matter or dark matter. The deceleration or acceleration of the expansion is governed by the sum of energy-density and pressure in the Universe. In Newtonian gravity, the force of gravity only depends on the mass, and since mass is always positive we always get an attractive force. But in Einstein's theory of gravity the curvature of space and time, which gives rise to the force of gravity, responds to the total energy which is composed of both rest-mass energy and thermal (or pressure) energy. While the rest-mass energy of particles is positive, there is no fundamental reason why the pressure should be positive. Hence a negative pressure can give rise to repulsion and hence acceleration of the expansion of the Universe.

How should we think of negative pressure and its effect? A simple example is quite easy to come by. Imagine a solid airtight tube with a piston in it. If the pressure of the air within the tube is higher than the outside atmosphere we would have to put some effort into pushing the piston in. The amount of energy is proportional to the pressure. Compared to the outside atmosphere the pressure in the tube is positive.

But if the pressure inside the tube is less than the atmosphere, we would have to put effort into pulling the piston out. The energy to do this is again proportional to the pressure in the tube, which is now negative compared to the atmosphere, but as we are pulling the piston out the energy is also negative compared to the energy we used in the positive pressure case. Hence we can have negative pressure giving rise to negative energy which will influence the curvature of spacetime. Of course this analogy breaks down when we try and think of a Universe filled with a negative pressure substance, where there is no 'outside'.

All of the observed effects can be understood by the presence of a negative pressure substance dominating the total energy budget of the Universe. The major problem with this conclusion is that there is no known fundamental substance with this property. This leaves us with two logical conclusions. Either we are now seeing the effects of some new substance not previously seen before, or something is wrong with the theoretical picture which brought us here.

The constancy between the different observational routes we can take to come to the same conclusion argues that we have not made a simple mistake. As well as the CMB galaxy redshift and supernova observations we now have many other observations leading to the same results, so disregarding one, or a number of results, still leads us to a negative-pressure problem.

But perhaps one of our more basic assumptions is wrong? The evidence leading to dark matter is independent of any consideration about the acceleration, however given we now have two unknowns it is tempting to think there may be some connection. But perhaps more worrying is that we have to rely on Einstein's theory of gravity at all stages to reach our conclusions. Perhaps the problem is here? This is a very interesting possibility and one we should be open to, especially as gravity has not been well tested on these large scales in the Universe.

Collectively, all of these models to explain the dominant energy, negative-pressure problem have been called 'dark-energy' models, in analogy with dark matter. (This can be confusing since Special Relativity tells us that matter is the same as energy, but generally dark matter and dark energy are not supposed to be the same.) The dark-energy problem and its solution has become one of the major problems in not only cosmology, but in all of physics. In Section 1.8 we discuss in more detail the ideas being proposed to solve the dark-energy problem.

But even if we have a number of possible solutions to the dark-energy problem, how can we distinguish between them? In general we have two generic methods of probing the dark energy. The first is to study the expansion history of the Universe. This was the method employed by the supernova studies, where we learn about the expansion history through the dimming of light with the distance it has travelled over the lifetime of the Universe. Both distance and time depend on the expansion history of the Universe.

As well as using the dimming of light we can also use the geometry of space to measure dark energy. We used the relationship between the apparent angular size of a known distance on the CMB sky to measure the global geometry of the Universe in Section 1.6. This was really measuring a fixed triangle between one point at us, and two points at the time the CMB formed. If we repeated this at different distances we could use this to again measure the expansion history of the Universe. While we only have one CMB sky, we can instead use the distance set by the baryonic wiggles seen in the galaxy power spectrum discussed in Section 1.6. The wiggles in the galaxy power spectra are the same as those seen in the CMB power spectra, but can be traced in the galaxy distribution over time. Since light travels at a finite speed, as we look further away we see a younger universe, and can measure the baryon wiggles at different distances and time. Hence we can measure the geometry and history of the expansion.

The second way to probe the dark energy is through its effect on the dynamics of the large-scale structure in the Universe. As the expansion of the Universe begins to accelerate, this will inhibit the gravitational collapse of structure. Eventually we can expect the gravitational formation of clusters of galaxies, and the formation of galaxies themselves, to halt. Since the formation of structure is slowing down now, and we know how much structure there was at the time that the CMB formed, there must have been more structure in the recent past, compared to a matter-only model. By comparing the history of the growth of cosmological structure with cosmic time we have an independent way to probe the history of the expansion and hence probe dark energy. Since both the formation of structure and the measurement of geometry probe the expansion history, we should expect them to agree.

We can exploit the dark energy's effect on the dynamics of structure and geometry of the Universe using our three familiar probes: the

BAOs in galaxy redshift surveys, supernovae Type Ia and the CMB. The intrinsic CMB does not tell us very much about the dark energy. Indeed, its main contribution has been to tell us that the Universe has a critical energy-density. But in finer detail, the photons travelling to us from the CMB are affected by the structure that they pass through. The most significant effect is due to the slowing down of structure growth. As a result of this change, photons which enter the gravitational field of a clump of matter will find it has a smaller gravitational field when it leaves it, compared again to a matter-only Universe. This effect is called the Integrated Sachs–Wolfe (ISW) effect, and it has already been detected.

Another probe of the dynamics of the growth of structure comes from looking at the number and evolution of galaxy clusters. As we noted, if the growth of structure is slowing down now, there must have still been a lot of structure around in the recent past. This implies that in the recent past the number of clusters of galaxies should be similar to today. In a non-accelerating universe, these clusters of galaxies would have only just formed and so we would not see any in the recent past. Just by counting the abundance of galaxy clusters, and their abundance with redshift, we can probe the dynamics.

Yet another, and potentially very important, probe of the geometry and dynamics of the Universe comes from gravitational lensing. As light travels across the Universe from distant galaxies it gets scattered by the gravitational potential of matter. This effect was first predicted by Einstein in his General Theory of Relativity, and beautifully confirmed by the British astronomer Sir Arthur Eddington in 1917, who measured the predicted bending of light by the Sun. Since then, lensing by larger and larger structures has also been measured. The bending of light around galaxies produced spectacular 'Einstein rings', while the gravitational field of galaxy clusters can heavily distort the image of more distant galaxies.

Even larger structures distort the images of distant galaxies, but because the variations in the gravitational field are so small, the distortions are also small, typically only a 1% distortion. A useful analogy

FIG. I.10 An example of gravitational lensing around a cluster of galaxies by the Hubble Space Telescope. The bright sources are the main cluster galaxies, while smaller, fainter images are more distant galaxies in the background. The large arced images are distant galaxies whose images are heavily distorted by the gravitational field of the foreground cluster. Source: hubblesite.org.

is the more familiar glass optics. The effect of a galaxy cluster is very much like holding a glass ball in front of a background pattern. The size of the distortion can be used to infer the strength of the lens – in our case the mass it contains. Gravitational or 'weak' lensing (WL) by the large-scale distribution of matter is analogous to looking through an irregular glass panel. At any individual point it may be hard to see an effect, but by looking at the whole glass plate one can see an induced irregular pattern.

Even so, these small distortions have been detected and used to measure the size of the variations in the gravitational field. Yet again these agree with the Standard Cosmological Model.

Gravitational lensing is unique in that it can probe both the geometry and dynamics. The lensing effect depends both on the distribution of matter and its gravitational field, and the geometry of the observer–lens–source configuration. By comparing the lens distortions at different positions on the sky and redshifts, one can probe both the geometry and growth of structure in the Universe. Indeed, one can go further and separate out the geometric effects of the Universe from the growth of structure and study them independently.

This combination of five basic probes of dark energy, SNIa, CMB, BAOs, clusters and WL provides us with a series of tests of dark energy which can be compared and combined in different combinations to test for consistency between the measurements. These consistency tests will prove a very strong test of the dark-energy picture. In addition, each of these methods for probing the dark energy depends on the geometry of the Universe and the growth of structure in different ways. Any measurable differences between different approaches could help distinguish between competing models of the dark energy (see below).

Along with the serious problem of the energy-scale of the dark energy there is also a mysterious timing problem. Until recently the energy-density of the Universe has been dominated by first radiation and subsequently by matter. During this time the dark energy would have had no noticeable effect on the evolution of the Universe, and so would have been very hard to detect by any observers. In the future, in the simplest models of the dark energy, the expansion of the Universe will accelerate away putting a stop to the formation of structure. If galaxies stop forming we can expect the stars will burn out their nuclear fuel and die, leaving no habitable conditions for life.

Outstanding problems for the Standard Cosmological Model

The Standard Model of cosmology, with only a handful of the parameters (six in its simplest form), seems to account for all of the observations we have. As we shall see this is an awesome achievement for a model, since there are millions of individual 'data points' to be compared with. But like only the best models, while the Standard Cosmological Model does a fantastic job of explaining the observations, it also provides us with a new set of very fundamental problems to understand and explain. And these problems seem to go to the heart of physics.

The first outstanding problem in cosmology is where did all of the normal matter in the Universe, loosely referred to as baryons by cosmologists, that makes up the galaxies, stars, planets and you and me, come from? This process, called 'baryogenesis', should be explained by a particle physics process in the early Universe, but so far the actual mechanism has eluded physicists.

The second outstanding problem is how did the galaxies form? This is very much an astrophysical problem. We already seem to have a good idea that the initial conditions for structure in the Universe may have come from inflation, and so we can assume galaxies arose on the site of dark-matter haloes, and very good observations of galaxies themselves today. But how galaxies actually formed from a diffuse cloud of gas into a well-structured spiral of gas and stars, or amorphous elliptical ball of stars, is still poorly understood.

The third problem is to find out if inflation actually happened. So far all of the evidence has been circumstantial. What is needed is a real 'smoking gun', a prediction which is then confirmed.

The fourth problem is the question of what is the dark matter. We assume it is a particle, and there is observational support for this, so we also assume it will one day be detected in the laboratory, and become a part of particle physics. But until then our main source of information about dark matter comes from cosmology.

The fifth and final outstanding problem is by far the most mysterious: What is the dark energy?

1.8 SOME THEORETICAL IDEAS FOR THE NATURE OF DARK ENERGY

While this essay has primarily been concerned with the observational evidence for the Standard Cosmological Model within a consensual theoretical framework, the unexpected appearance of a dark-energy component to the Universe poses a major theoretical challenge. As we shall see, the most obvious explanation in terms of vacuum quantum fluctuations yields nonsense. At such a time we can expect huge upheavals in the foundations of physics as we try to understand why. Here I discuss some of the currently topical theoretical proposals to deal with the dark-energy problem.

Dark energy from quantum gravity

One possible origin for the dark energy is that it is due to vacuum fluctuations. One of the basic principles of quantum mechanics, which govern the behaviour of forces and matter on small scales, is the Heisenberg 'uncertainty principle'. This states that if we know the position of a particle now we will have no knowledge of where it will be next. Even the number of particles in some volume cannot be known completely accurately. This implies that even if we try and remove all the particles in some volume to make a perfect vacuum, we cannot as there is always the chance that some particles will appear spontaneously. The appearance of these particles is due to an idea suggested by the British physicist Paul Dirac who proposed the idea of antiparticles, particles with the same mass but opposite charge to normal particles. For every particle there is an antiparticle twin. Pairs of particles and antiparticles can be created spontaneously due to the uncertainty principle and annihilate in a time too short to detect. The energy in the vacuum is given by how rapidly these particle–antiparticle pairs are being produced. The existence of vacuum energy has been proven by an experiment to detect the Casimir effect (see Section 1.4). Vacuum energy could have the right properties to be the dark energy.

The problem with all of this is that the energy of the vacuum increases on shorter and shorter distances. As we approach infinitely small distances, the vacuum energy diverges to infinity. Particle physicists have circumvented this problem, as laboratory experiments can only measure properties with respect to the vacuum. So the actual energy in the vacuum itself cannot be measured in a laboratory. But on cosmic scales, all energies contribute to the curvature of spacetime, including vacuum energy. If we add up all the contributions to the vacuum energy, a naïve estimate which sums over all scales diverges to infinite. This is clearly bad news.

We can revise the estimate with the following argument: our understanding of gravity and matter must break down at very small scales. After all, quantum effects become more important on smaller scales, while gravity is more important on large scales. At some point the two must meet. This could happen at a scale of 10^{-35} m, at the so-called Planck length scale. If we only sum up the contribution to the vacuum energy down to this scale we get a finite answer. However, this is still 10^{120} orders of magnitude larger than the observed energy scale. Earlier on we noted that there was no contradiction between particle physics and cosmology. But here is an outstanding contradiction!

Perhaps we are still too extreme. What happens if we try to find the length scale which would give us the right magnitude we are looking for? In that case we must cut the length-scale off at a scale that is well within the range of particle accelerators, and no mysterious effects have been observed here.

The discrepant nature of the observed dark-energy effect and our particle physics order-of-magnitude estimate is perhaps the biggest problem to explain. There seems to be no easy solution. In fact the only proposed solutions require us to make a huge leap of faith about the nature of reality.

Quantum quintessence

While the solution to the deep problem of the size of the dark-energy effect may require a substantial change in our understanding of nature, this does not preclude us from studying simpler models for how the dark energy may arise, exploring them and testing them against data. One way to make a fundamental model of the dark energy is to assume the dark energy is a new force which can be described by the same laws used in particle physics. The simplest possible type of particle-physics model which can be adapted for dark energy is to assume the new force has only an amplitude at each point in space, and does not depend on direction. This type of force is modelled by what is called a scalar field, but in this context such models are called quintessence (or fifth element) fields. Requiring this new field to conserve energy is enough to tell us how this field will evolve, given other assumptions about how the field interacts with itself.

In the simplest versions of these models the pressure depends on the details of the scalar-field interactions, but is bounded such that if the forces between the scalar-field particles are everywhere zero, the scalar field looks just like a cosmological constant. Introducing forces can only decrease the pressure. Hence if observations indicate that the pressure is more negative than a cosmological constant, we can rule out all of this type of model.

Of course, theorists are ingenious and have found ways of creating non-standard models with more negative pressure. The price of doing this is the introduction of non-standard ways of making models which are not based on known laws of physics. Despite the difficulty of realising a dark-energy solution with more negative pressure than the cosmological constant we can still take a more phenomenological approach and assume they may exist. These models have some interesting new features, the most important of which is that their energy-density will increase with time. These models are collectively called 'phantom dark-energy' models. This behaviour is different from the cosmological constant, whose energy-density of course stays fixed, and the more general quintessence models whose energy-density tends to decrease with time.

The increase of energy-density in the phantom dark-energy models tends to be a run-away process which ends when the scale factor of the Universe becomes infinite. Leading up to this the expansion rate becomes infinitely fast, which will pull apart clusters, galaxies, planetary systems, stars and planets and will eventually pull apart atoms and perhaps fundamental particles. This scenario is called the 'Big Rip'. Of course the Big-Rip solutions are idealised, just as the Big-Bang and 'Big-Crunch' events are idealised. In practice our knowledge of physics will break down prior to these events and we are unable to predict what will happen next. Even though the Big-Rip scenario seems difficult to realise, it does open up some interesting new possibilities.

One problem with these models based on extrapolating particle physics is that the size of the dark-energy effect is unexplained. We are merely proposing mechanisms by which the accelerated expansion may work. The size of the effect has to be 'fine-tuned', or put in by hand, and so these models are ultimately superficial explanations for dark energy.

Extra dimensions

Instead of assuming that the dark energy is due to some new force, another logical explanation might be that we do not have the correct model for gravity and Einstein's theory is wrong or incomplete. This is perhaps not so unlikely. We have only tested gravity to high accuracy within our solar system and in systems like binary stars. Perhaps on larger scales there are other effects that change the laws of gravity which we are only now seeing.

One way to open up more possibilities for gravity is to relax the assumption that the number of spatial dimensions is three. There is some motivation for this from superstring theory. Superstring theory tries to explain the existence and properties of all particles by assuming they are all in fact different manifestations of harmonic oscillations of fundamental strings, or sheets or higher-dimensional objects. These models only make logical sense if they exist in a 10-dimensional space. But we only see three spatial and one time dimensions. To reconcile this superstring theorists used a trick introduced in the 1920s by the physicists Oskar Klein and Theodor Kaluza, who argued the number of dimensions could be greater than the observed three if you could make the extra ones small by rolling them up.

While these six extra dimensions have to hidden away to avoid interfering with reproducing our observed four spacetime dimensions, the possibility of a larger extra dimension appeared as a consequence of unifying the known superstring theories. In the mid 1990s the American physicists Edward Witten at the Institute for Advanced Studies in Princeton and Petr Horava at Rutgers University in New Jersey noticed that in string theory there were two types of string: open strings whose ends had to finish on a larger-dimensioned membrane which represented matter particles, and closed loops representing forces and in particular gravity. The unification implied Horava and Witten could construct a model where a membrane embedded in a larger-dimensional space represented our Universe where the matter particles were stuck to the membrane, or confined to our Universe. But the closed-loop gravity was free to travel in the higher dimension. This opened up the possibility of large extra dimensions where gravity had a different strength.

In 2005 Lisa Randall and Raman Sundrum showed that these extra dimensions need not be small or compact, so long as the distances along these extra dimensions were shrunk by a 'warp factor'. They showed that under these circumstances, gravity in the full 5-dimensional space was much stronger than in our reduced 4-d Universe. We only slice through gravity's full strength. The possibility that gravity may be a much stronger force may help with trying to unify gravity and the other, much stronger, particle forces. But if true, it would also change the predictions of the very early Universe when the energy-density of the Universe was much higher. Other effects are possible even in the low-energy Universe we see today. If the Universe is in motion through the extra dimension an accelerating force is produced within the Universe, creating the observed acceleration of the expansion. However, like the quintessence models, this requires fine-tuning to explain the size of the dark-energy effect.

Cyclic universes

The possibility of extra spatial dimensions has also given rise to the possibility of cyclic evolution of the Universe, and another possible explanation of dark energy. While an old idea in ancient cosmology, in relativistic models the idea dates back to Geoffrey Burbidge in the 1930s. The original idea was that the Universe expanded from an initial singularity, then collapsed towards another singularity where it bounced and restarted the whole thing again. This failed because either all its properties were set back to the same at the start, and so would just be one universe where we had identified the start and end singularities, or structures like black holes were carried through successive cycles. Given the number of black holes in our Universe seems to be the same as the number of galaxies this cycle can only have started recently.

With an extra dimension we can avoid this. In some superstring models we find not only one universe with matter fields confined to the membrane-universe, but two, separated by a large extra dimension. Building on Randall and Sundrum's toy model, in 2001 Justin Khoury, Paul Steinhard at Princeton, Burt Ovrut at the University of Pennsylvania and Neil Turok in Cambridge speculated that the space between these universes may be dynamic. They found that the forces generated by moving the two universes through the large extra dimension towards each other resulted in an accelerated expansion of the universe, again just as observed. In a very interesting twist, Steinhard and Turok suggested that when the universes eventually collide the kinetic energy of the moving universes would be turned into hot radiation in each universe which would then look like a Big Bang to someone in those universes. They called this model the Ekpyrotic Universe after the ancient Stoic cosmological model where the Universe is created in a burst of fire ('out of fire').

After the collision of the two universes they would then bounce away from each other until their attraction pulled them back together again. The collision of the universes would look to us like a Big Bang, while the long period of accelerated expansion preceding it would act like inflation, clearing the Universe of unwanted leftovers. This process of repeated collision and bounce of universes would allow an unlimited process of Big Bangs and cosmic acceleration.

While superstring theory provides a possible motivation for this scenario, there are other arguments for it. Most notable are the similar conclusions of a cyclic Universe reached by Roger Penrose based on the deep use of conformal transformations, which are presented later in this volume.

The multiverse

The idea that there may be multiple universes, or multiverses, is not new. However in recent years the possibility has been revived as a result of its ubiquitous appearance in a number of theories which we must take very seriously. One example is during a period of cosmological inflation. As we have discussed, inflation predicts there was a period of rapid expansion in the early Universe which set the initial conditions for the Big-Bang model. But to start inflation requires the right conditions: in particular that the energy in the inflation field driving inflation is large enough. In some pictures of inflation there are always parts of the initial Universe which will undergo inflation, seeding causally separate multiverses. This is the 'eternal inflation' scenario.

Another realisation of the multiverse is in superstring theory. Earlier we described how superstring theory implied six compact extra dimensions to allow for consistency and to unify superstring theories required yet another, larger, dimension with the possibility of another universe in it. This picture can be generalised to allow any number of universes moving around and colliding in this extra dimension.

In a further development of the superstring theory, it was suggested by Leonard Susskind at Stanford University in 2003 that the unification of the early superstring theories implied that there was a huge number of potential and possibly actual universes. The properties of these universes are predicted to vary, for example in their values of fundamental constants, providing a 'landscape' of possible universes. A possible consequence of this is an explanation for the observed value of the cosmological constant. As we have seen, quantum physics gets the size of the vacuum energy wildly wrong. But what if the prediction is correct, on average? In different universes the actual value could be different. This would suggest that if the value of the cosmological constant was usually much larger in most universes, we are most likely to live in a universe where the cosmological constant is just small enough for life to have evolved, providing an anthropic explanation for its observed value at the cost of accepting the multiverse.

Even before the 'superstring landscape' picture appeared, the Anthropic Principle had been used to understand the problem of the cosmological constant. In 1987 physicist and Nobel prize-winner Steven Weinberg used the same reasoning to suggest that if there was a large number of possible states the Universe and its fundamental constants could be found in then the fact we find ourselves in a Universe where the value is fantastically low is just a consequence of the fact that we would not exist in a Universe in which the value was higher. Weinberg took this idea further and suggested that given the most probable values for the cosmological constant were very much higher, we were most likely to live in a Universe where the cosmological constant just allowed life, and was very unlikely to be very much smaller. On this basis he boldly proposed the value of the cosmological constant would not be zero. This remarkable prediction has turned out to be true, but if we are to accept its reasoning we have to buy into a multiverse picture.

1.9 FINAL REMARKS

Cosmology has come a long way in the last eighty years, since Slipher' discovery of the expanding Universe, Hubble's discovery of the distance–velocity relation and Einstein and Friedmann's relativistic cosmologies. It is remarkable that the simplest cosmological models have survived so well with the establishment of the Big-Bang model, which in turn motivated inflationary cosmology to explain why the simplest model works, and finally the establishment of the Standard Cosmological Model with dark matter and dark energy which is able to explain all that we can see today. But like any good model we have some really fundamental issues still to tackle: Did inflation really happen? What is the dark matter? What is the dark energy? The answer to the dark-matter question may be answered soon if a dark-matter particle is created in the lab. But for the other two we must rely on cosmological observations and perhaps a leap of insight into the nature of space and time and quantum theory before we will find a satisfactory answer.

The references below are some suggestions for further reading.

BIBLIOGRAPHY

Caldwell, R. R. (2004) Dark energy, *Physics World*, May issue.

Susskind, L. (2005) *The Cosmic Landscape: String Theory and the Illusion of Intelligent Design*, Little, Brown.

Kirshner, R. P. (2004) *The Extravagant Universe: Exploding Stars, Dark Energy, and the Accelerating Cosmos*, Princeton Science Library.

2 Quantum spacetime and physical reality

Shahn Majid

2.1 INTRODUCTION

Whereas the previous chapter tells us about mysteries surrounding the physical structure of the Universe from a largely observational point of view, in this essay I will approach the problem of space and time from a theoretical point of view. This is about the conceptual structure of physics, why in fact our current concepts of space and time are fundamentally flawed and how they might be improved. I will explain in detail why I think that spacetime is fundamentally not a smooth continuum at the pre-subatomic level due to quantum-gravity effects and why a better *although still not final* picture is one where there are no points, where everything is done by algebra much as in quantum mechanics, what I therefore call 'quantum spacetime'.

The idea of 'moving around' in space in this theory is replaced by 'quantum symmetry' and I shall need to explain this to the reader.[†] Symmetry is the deepest of all notions in mathematics and what emerged in the last two decades is that this very concept is really part of something even more fundamental. Indeed, these quantum symmetries not only generalise our usual notion of symmetry but have a deep self-duality in their very definition in which the role of the composition of symmetry transformations and a new structure called a 'coproduct' is itself symmetric. For our purposes, quantum

[†] This is my own route to 'noncommutative geometry' but see the essay of Connes for a deeper one and a different application of it, namely to elementary particle physics by means of noncommutative 'extra dimensions' added onto usual spacetime (which remains commutative).

On Space and Time, ed. Shahn Majid. Published by Cambridge University Press.

symmetries are needed in order to extend Einstein's theory of Special Relativity to quantum spacetimes. In one resulting model, if you measure t (when you are) and then x, y, z (where you are) you may get a different answer than if you do it in the other order. The error may not be a lot, about 10^{-43} seconds, but it's nonzero and this means that the very notion of a point in spacetime is incorrect. Moreover, we shall see that some of the resulting quantum-gravity effects could in principle be tested by experiments today, which is an entirely healthy way to do physics.

If one asks what made Einstein's even more revolutionary 1915 theory of General Relativity possible, it was perhaps not only cutting-edge experimental possibilities which allowed Einstein's prediction that light would be bent by gravity to be tested, and the new mathematics of Riemannian geometry already invented by Riemann in 1865, but their conjunction with his deep philosophical insight. I will argue that the same is true today and that we can't actually understand what space or spacetime *truly is* without at the same time addressing bigger and much deeper questions such as, 'what is is?' (to misquote a former US president). Viewing quantum symmetries as a microcosm of the problem of quantum gravity, I will argue that a deep duality between product and coproduct provides a glimpse of such a new philosophy for quantum gravity and for the fundamental nature of space and time. In it, spacetime and quantum theory are in a dual relationship in which 'each represents the other', providing in the last third of the essay my own answer to the allegory of Plato's cave often discussed by philosophers.

Along the way, the essay provides at least some insight into such theoretical questions as '*why are things quantised*', '*why is there gravity*' and '*why is there a cosmological constant*'. The first two are usually taken for granted, but in my view a true philosophy of physics should help explain even these most basic of discoveries. The third by contrast is a major theoretical problem of our times which was discussed from the observational side in the preceding chapter as 'dark energy'.

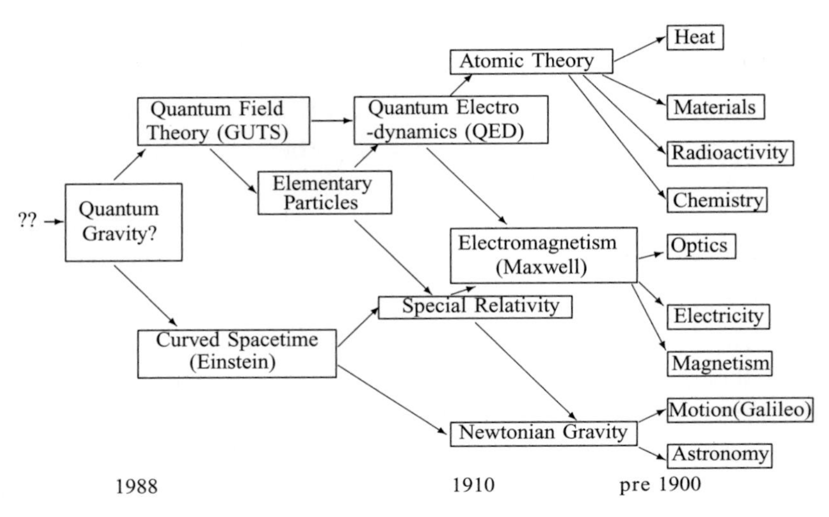

FIG. 2.1 Schematic of evolution of theoretical physics. Arrows indicate how the laws of one area may be deduced as a limit of more abstract ones, from Majid (1988a).

2.2 A HOLE AT THE HEART OF SCIENCE

It's the things that we most take for granted that have the tendency to come back and bite us when it really matters. The nature of space and time is generally taken for granted. But our assumptions about them seem to be inconsistent and as a result, if we are honest, theoretical physics is currently derailed at its very core. Let me explain.

The first thing that the reader needs to understand is about how science works. Otherwise, how is it possible to have a major 'hole in science' like this without the whole edifice collapsing? This is a matter of approximations, and indeed knowing what approximations to make in a given situation, how to extract the science, is the art of physics that is hardest to teach. On the one hand most scientists (and we will too) assume an irreducible quality to our Universe, everything affects everything else. At the same time any branch of science is about focussing in on some particular mainly self-contained paradigm where much else can be treated as negligible. This is depicted in Figure 2.1 and goes hand in hand with the way scientific fields develop in practice from experimental observations. Any experimental observation or effect is within some range of forces or scale or context and this sets the scene for what to count and what to ignore.

One branch of this 'tree of physics' is quantum theory. Our story here is from the turn of the nineteenth to twentieth century. Newton had already shown how the motion of the heavenly bodies could be explained in terms of calculus and simple laws that apply to all macroscopic particles (such as apples) and presumably to particles of any size. This had led up until the end of the nineteenth century to classical or 'Newtonian' mechanics for the motion of particles. However, Newton, Huygens and others also did fundamental work on light or the theory of optics. It was understood that light was a wave much like a water wave, capable of reflection and refraction, and eventually of interference. But by the late nineteenth century physical observations about 'black-body radiation' began to appear not quite consistent with this. Normal waves after all should have a continuous range of energies but this gave an incorrect prediction for the radiation energy spread and what the physicist Max Planck suggested instead in 1900 was that in fact the possible energy should be discrete or 'quantised' in lumps, which later came to be called 'photons', each photon having energy

$$E = h\nu \tag{2.1}$$

where ν is the frequency at which the light vibrates (its colour) and h is Planck's constant of proportionality needed for this (its value is 6.6×10^{-34} joule-seconds in scientific units). Einstein contributed here with an explanation of a certain photoelectric effect in terms of a photon hypothesis. And so the seeds of quantum theory were sown. Einstein also provided in his 1905 theory of Special Relativity a famous formula $E = mc^2$ for conversion between mass and energy, where c is the speed of light (about 300 million metres per second). We will use this formula as the definition of the equivalent 'mass-energy' m of a photon of light when comparing with other particles (photons of light are 'pure energy' particles in the sense that if one could bring a photon to rest it would be massless, on the other hand it is impossible to bring a photon to rest as they always travel at the speed of light so this is a moot point). It is also a good idea to think of waves in terms of their wavelength λ (the distance between peaks of a wave) rather than their frequency of vibration. The conversion between the two is the wave speed so for light the relationship is $\nu = c/\lambda$. Putting these two observations into the above Planck's relation for light one has

$$\lambda = \frac{h}{mc}, \tag{2.2}$$

a formula that we shall need below. The nice thing is that this formula also applies to other particles and indeed it was first proposed by the English physicist Arthur Compton as a measure of the localisation of an electron. Here λ defined by (2.2) is called the Compton wavelength and in this context we refer to (2.2) as Compton's formula.

This was perhaps the first hint of matter particles also being waves, an idea particularly developed by a young student Louis de Broglie. He explicitly proposed that if light waves were also particles, maybe fundamental particles like electrons were also waves. This turned out to be right, for example Figure 2.2 shows some electron waves breaking around two atom-sized defects on the surface of a copper crystal (taken with a scanning tunnelling microscope). De Broglie and Compton both proposed a famous formula

$$\lambda = \frac{h}{p} \tag{2.3}$$

as a relativistic counterpart to Planck's relation (2.1) and now to hold for any kind of particle. Here λ is the wavelength of the wave and p is the impulse felt when a particle of this type hits an object (this is called its *momentum*). For a slowly moving massive particle its momentum is the product of its mass and its velocity (so a heavy object hitting you slowly can have the same impact as a lighter object hitting you more quickly) while for light it is similarly mc with m the mass-energy as above. Putting in $p = mc$ recovers our previous formula (2.2) in the case of light, while doing the same for matter waves is not really correct but defines the Compton wavelength as a

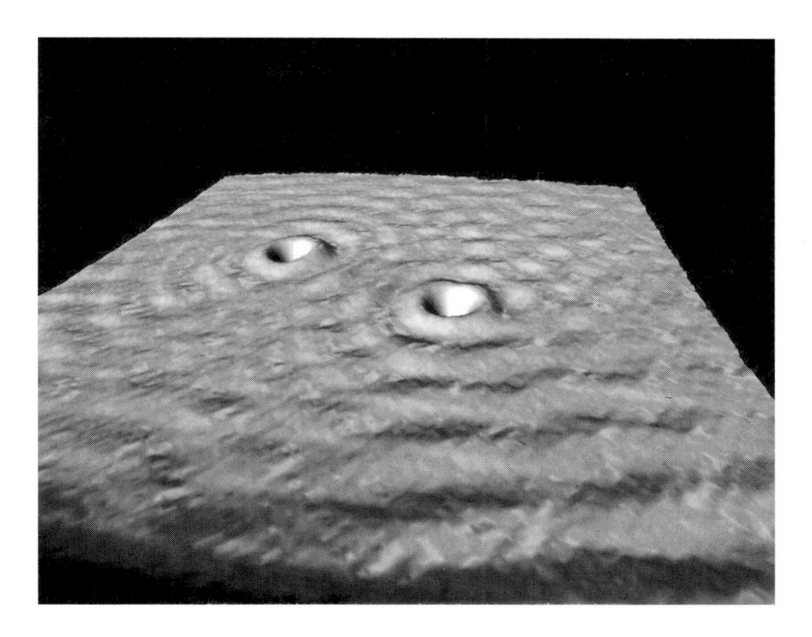

FIG. 2.2 Electron wave scattering on atomic defects on surface of a copper crystal. Courtesy, Don Eigler, IBM.

rough measure of the degree of localisation of the particle due to its wave-like aspect in relation to its mass.

What is important for us is that smaller wavelengths correspond to larger masses. For example, BBC Radio 4 transmits on a radio frequency (radio waves are a form of light but not visible to the human eye) of 93.5 million vibrations per second or a wavelength of 3.2 metres and each photon of it has a mass-energy of 0.69×10^{-39} g. Microwaves are in the centimetre range with correspondingly more mass-energy. Visible green light (for example) vibrates at about 550 million million vibrations per second so has a wavelength of 0.55 micrometres and each photon of it has a mass-energy of 4×10^{-33} g. X-rays are more energetic photons, in the nanometre range (a nanometre is one American billionth of a metre). 'Gamma ray' is the name given to the most energetic photons, with wavelengths smaller than 30 picometres (a picometre is one million-millionth of a metre). Meanwhile, an electron has a mass of 9.1×10^{-28} g hence a Compton wavelength of 2.43 picometres. Protons which make up the nuclei of atoms have a mass of 1.7×10^{-24} g, so about two thousand times smaller. Some of these numbers are plotted a bit later on, in Figure 2.5. Now what you should note about all this is that if you try to use waves or matter to probe the structure of space and time (and how else are you going to do it?) you will never be able to resolve things more sharply than the wavelength of the particle. That is why an electron microscope is so much sharper than an optical microscope while radio waves are more useful than telescopes looking at a large scale in astronomy. So far so good, one has an enormously accurate and versatile theory, quantum theory, that describes both matter such as electrons and protons, and force-particles such as photons, but which does not describe gravity.

Now wind back to another aspect of Newtonian mechanics, namely gravity. This is something equally tangible and in many ways more tangible. One of the things that greatly bothered Newton about his own theory was the idea that a bit of matter over here exerts a gravitational force of attraction on a bit of matter over there. He provided a formula for this and it worked pretty well, but still it smelled of witchcraft. How could something in one place instantly affect something in another place? Einstein's answer in his 1915 theory of General Relativity (GR), which was motivated by philosophy but which also provided small corrections to Newton's gravity that were subsequently verified, was that gravity is actually a curvature of space and time. He also provided an equation, now called 'Einstein's equation', to express this. In this theory each particle makes space and time bend a bit, and this makes other particles bend towards it. The standard mental image for curved space is to think of a rubber sheet with a heavy marble placed on it, making a dip. Other particles moving on the resulting curved surface would tend to roll down into the dip (as well as make their own dips) and this is gravity. There are a lot of things wrong with this image; the most important is that there is no external force making things fall into the dips as in the rubber model. It is the geometry of the dip itself that makes other particles

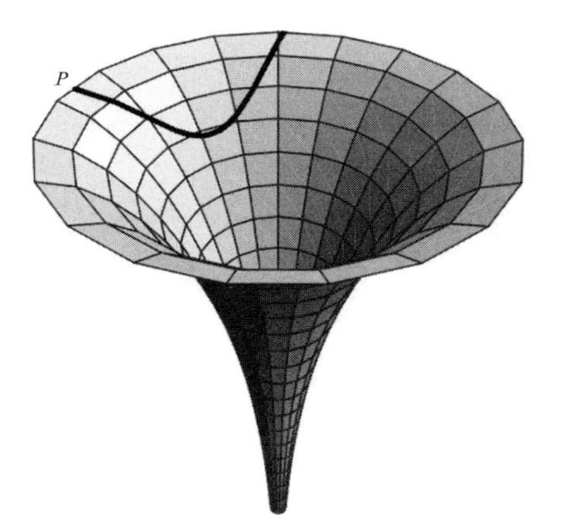

FIG. 2.3 Surface of constant negative curvature. Ants crawling at point P towards the depression but following their shortest path would be deflected *outwards*. This is because the angles of a triangle of such geodesics add up to less than 180 degrees.

bend into it without the need for any exterior explanation. This is a key point for which a better analogy (also well-worn) is an ant moving on the surface. Let's say that the ant always wants to move along what for it is a straight line, getting the shortest distance from its starting point to wherever it is going. This is actually something that the ant can determine at each stage of its journey from the local geometry and without knowing its final destination. Such a path is called a *geodesic* and how such geodesics behave captures the true geometry of the surface. For example, when three geodesics form a triangle their angles add up to 180 degrees on a flat surface. Positive curvature is when the angles add up to greater than 180, and negative curvature is when the angles add up to less. An example of a geodesic on a surface of constant *negative* curvature is shown in Figure 2.3. Any geodesic triangle here would have a sum of angles less than 180 degrees, which tells you that a particle moving by in the vicinity will be deflected *away*. Try the same thing on a sphere (where a geodesic is a great circle) and you will get a sum greater than 180 degrees. There is still something wrong with this example as regards gravity: it is in a 2-dimensional spatial surface rather than a 4-dimensional spacetime surface but the ideas are similar. In the correct (Lorentzian) notion of distance for spacetime, a particle moving in the vicinity of a well such as in the figure would be deflected inwards, and this is gravity.

FIG. 2.4 Galaxy NGC 4696 in the Centaurus Galaxy Cluster glows because its centre is one huge black hole swallowing up stars, which emit radiation as they fall in. Credit: NASA/CXC/KIPAC/S. Allen *et al.*, X-ray; NRAO/VLA/G.Taylor, radio; NASA/ESA/McMaster Univ./W. Harris, infrared.

We do not need to understand curved spacetime in detail here,† but we need one fundamental formula about black holes. We recall that a key prediction of Einstein's theory, by now believed by astronomers to be well-confirmed, is that when enough mass is compressed into a small volume it forms a black hole. The gravitational field here is so strong that nothing, not even light, can escape from inside the black hole. In modern language, spacetime curves in on itself to form a trapped sphere. The size r of the black hole, which it is convenient to define as $\frac{1}{2}$ the radius in a standard description of the sphere from inside of which light cannot escape (the 'event horizon'), is given by

$$r = \frac{GM}{c^2}. \qquad (2.4)$$

Here $G = 6.7 \times 10^{-8}$ cm^3/g s^2 is Newton's constant that sets the scale for how much curvature is caused by the mass M inside. Once again,

† See the article of Penrose in this volume for more detail, in particular the Riemann–Christoffel tensor **R** encodes the curvature.

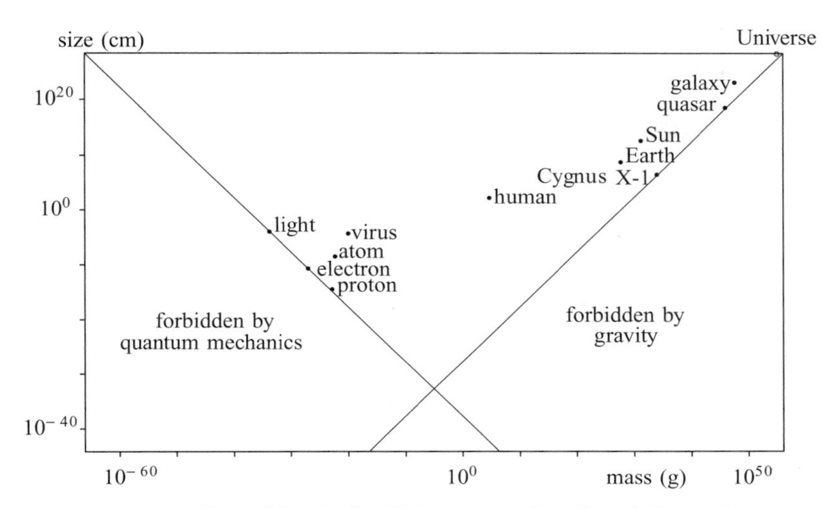

FIG. 2.5 Everything in the Universe on a log plot of size against mass-energy, from Majid (1988a). *Did we box ourselves in?*

we have an elegant and accurate theory (as far as we know) that applies to matter and the structure of space and time but this time does not take into account any quantum effects. Indeed, you would need to have a very tiny black hole to be concerned about quantum effects whereas astronomers who have looked for such things are more interested in ones ranging from something a bit larger than our Sun (such as Cygnus X-1, the first object more or less agreed to be a black hole out there) to giant black holes thought to be swallowing up the centre of many galaxies including our own. Another feature of black holes is that they are in some sense totally smooth on their surface and hence in many ways behave like fundamental objects of some kind.

But what about very very tiny black holes where quantum effects cannot be ignored or very very energetic massive particles where gravitational effects cannot be ignored? To study this question, I have summarised our discussion so far in Figure 2.5, where I indicate the mass-energy (understood broadly) and size of everything in the Universe on a log scale. The log scale here means that each notch on the axis is a factor of 10 American billion (10^{10}). Everything to the left is forbidden as shown by quantum theory in the sense that smaller masses or lower energy photons have larger size in the sense of wavelength. This was Compton's formula (2.2) explained above. Fundamental particles are governed by this formula and lie on this line, which we could view as the line dictated by quantum theory alone. Meanwhile, everything to the right is forbidden by gravity in the sense that too much mass in a given space forms a black hole and adding mass only makes the black hole bigger, which is the line dictated by gravity according to our equation (2.4). The intersection of the two constraint lines is the *Planck scale,*

$$m_{\text{Planck}} = 2.18 \times 10^{-5}\,\text{g}, \quad l_{\text{Planck}} = 1.62 \times 10^{-33}\,\text{cm} \tag{2.5}$$

approaching which we would need a theory of quantum gravity to understand. What is happening here is that as you try to probe smaller and smaller length scales using more and more massive quantum particles (going down the left slope) you eventually need particles so big that they form black holes (because we approach the right slope) and mess up the very geometry you are trying to measure. What this means is that *distances less than* 10^{-33} *cm are intrinsically unknowable.* From a logical positivist point of view they have no place in physics. This should already be a warning to us. Certainly, to resolve such tiny length scales approaching 10^{-33} cm, you need both quantum theory and gravity in a single theory to understand how to proceed.

Alternatively, consider lighter and lighter black holes, which means smaller and smaller ones (going down the right slope). By the time black holes are only about 10^{-5} g you would start to need quantum theory as well as Einstein's theory to describe them (because you approach the left slope). These are not merely esoteric questions. For example, it is believed that black holes very slowly evaporate due to Bekenstein–Hawking radiation.[†] If so, they slowly shrink. Eventually they reach 10^{-5} g and we have no idea what happens to them after

† This comes from thinking about quantum vacuum fluctuations for particles *in* the black-hole spacetime not from the black hole itself being quantised.

that. Many scientists, notably Hawking, have argued that black holes evaporate completely[†] but it would be better to say that at this size they are so small that they are quantum objects as well as curved in on themselves and as a result we do not actually know. More likely in my opinion is that they form stable hybrid objects in which the rate of evaporation is balanced by a quantum tendency according to (2.2) to have a larger size or wavelength if they were to become less massive. If so, the Universe could be littered with a dust of such black-hole remnants describable only by quantum gravity and perhaps this has something to do with the dark matter observed by astronomers. Such an explanation is thought not to tally with the expected number of black holes produced in current cosmological models, which appears to be far too small, but it is an idea to keep in mind. Again, we would need a theory of quantum gravity to know what kind of object we have as a remnant. It would be something which is an elementary particle and a black hole at the same time.

Now, the problem is that when theories develop in different parts of physics there is no assurance that they are going to be compatible where they overlap. Where such overlaps have happened in the past, both paradigms have been refined, maybe slightly corrected, and this led to a single better theory. This has convinced theoretical physicists of the irreducible nature of the Universe and of the success of the reductionist assumption that we can progressively boil down our knowledge to more and more encompassing theories and ultimately one all-encompassing theory of physics. This, however, is a matter of faith and, so far, it is exactly what has not been achieved for quantum theory and gravity. The fundamental problem at the moment is as follows. Going back to Newton's complaint about his own theory, Einstein's resolution to it was that gravity is not after all instantaneous. When a particle moves about its gravitational effects have to propagate, at the speed of light, as ripples in the fabric of spacetime. The existence of such gravitational waves is not really in doubt

[†] This features in Penrose's essay in this volume.

although it is only this year that experimenters hope conclusively to see them. So there is no need for witchcraft. Particles communicate the force of gravity by sending signals in the form of gravitational waves. The problem is that when these waves have a very small wavelength they should surely be governed by quantum theory just as other waves are. One would expect that there are quantum particles called (if they exist) *gravitons* just as electric and magnetic forces are mediated by photons. The problem comes when you try to develop the quantum theory of such gravitons. It may be that gravity is different from other forces such as electricity and magnetism, but then how else to reconcile it with other particles and matter which are described by quantum theory? If we do assume that there are quantum particles called gravitons, the problem, however, is much more severe than that nobody has yet seen one. Attempts to develop a quantum theory for gravitons have all turned out to be too sick in one form or another. These sicknesses are technical to explain and most likely (in my view) a result of wrong-headed methodology about quantum theory itself. But basically the problems are to do with severe infinities in the quantum theory that have their origin in the assumption of a continuum for space and time. Roughly speaking, if space and time are infinitely divisible (in the sense of a continuum) the contributions to the usual computation for the probability of physical processes coming from gravitons of infinitely short wavelength are so large that they result in infinities that cannot be made sense of. All of this probably has to do with the special role of spacetime in the quantum physics of matter and other particles. These all move *in* spacetime while gravitons have to *be* spacetime as well as move in it. Their double role appears too much for current theoretical ideas to cope with in any convincing manner. This is the fundamental hole in science. In summary:

(i) We have a conceptual inconsistency in science that we assume a continuum but in fact distances of less than 10^{-33} cm are intrinsically unknowable even within the theory; any probe of them would inevitably distort the geometry it was trying to probe beyond recognition.

(ii) If one ignores this one finds that the usual ideas of quantum theory fail when applied to gravity. Hence there is no conventional theory of quantum gravity.
(iii) This hole in our fundamental understanding of space and time does not stop physics working approximately in paradigms far from the Planck scale.
(iv) The Planck scale is the 'corner' where two different extrapolations from everyday scales intersect in an inconsistent manner.

Simply put, we assumed a continuum because that was the appropriate mathematics at everyday scales on Earth when modern science put down its roots, and we extrapolated this to large and small scales. But this turns out both to be illogical and to get us into trouble. This problem has been around for some decades now and never really addressed. It means that we do not understand for example what exists at the very centre of any black hole – classical general relativity says that the gravitational forces become infinitely strong at the very centre, but most likely quantum effects will change that conclusion. We also do not know what is left as black holes evaporate as mentioned above. We also do no know what happened at the 'Big Bang'. Again, classical general relativity and astrophysics suggest that the Universe is expanding or conversely that going back in time it was smaller and smaller. There is much more about that in the essay by Taylor in this volume. But at the point when the Universe was around 10^{-33} cm our theoretical understanding breaks down and theoretical physics cannot therefore answer the 'creation question'.[†] My own suspicion is that due to quantum effects there is no space and time 'before' this point; both space and time coalesced out of whatever was quantum gravity. Properly to understand all these and many other Planck-scale effects we would need a theory of quantum gravity and, as we have seen above, this would likely entail a completely new concept of space and time.

What is also interesting in Figure 2.5 is that life, our macroscopic scales, lie somewhat in the middle. The usual 'explanation' is

[†] But see the essay of Penrose, where it is argued that one can still say some things about it.

that things become simpler on the above two boundaries as phenomena are close to being forbidden hence tightly controlled there; hence conversely they should get more complicated as one approaches the centre of the triangle and hence this is most likely where life would develop. However, my view is that *we are in the middle because we in some sense built the edifice of physics around ourselves starting with things we could easily perceive and our assumptions in doing this have now 'boxed us in' and not without some inconsistencies at the corners.* We shall return to this philosophical point in Sections 2.5 and 2.6.

Finally, because this is a speculative article, let me make one more speculation from Figure 2.5. In this essay we will later speak of a duality-symmetry between the position of a particle and its momentum. This was particularly stressed by the philosopher Max Born and is sometimes called 'Born reciprocity'. We shall argue that it has a much deeper origin as a kind of duality between quantum theory and gravity or if you like a kind of reflection-symmetry between the two slopes in Figure 2.5, and that this duality is a guide as to what quantum gravity should be. For the moment I want to say that if there is a smallest position scale 10^{-33} cm then by Born reciprocity there *should* be a smallest momentum scale also, and this is indeed the case. Since the Universe is thought to have a finite size as we see in the figure, currently of the order of 3×10^{28} cm, this is also the maximum possible wavelength, hence according to de Broglie's formula (2.3) there is indeed a minimum momentum for a particle,

$$p_{\rm min} \approx 2 \times 10^{-55}\ \text{g cm/s}.$$

So the idea is that this should correspond to the Planck length scale in some other, dual, picture of physics where position and momentum are interchanged. If the dual theory was exactly equivalent and if we understood the full details of this equivalence, we could perhaps turn the argument around to predict the size of the Universe in the current epoch from the Planck scale in the dual picture. Such speculation is admittedly a bit alarming since it suggests that the fundamental constants that determine the Planck scale in the dual theory, and hence presumably our own fundamental constants, vary with time, given that our Universe is thought to be expanding. However, all that we need from these ideas at the moment is the observation that momentum, like position above, cannot be resolved below a certain momentum scale $p_{\min}$ and this time due to the finite size of the Universe.†

Now, if we divide the momentum by the speed of light according to the rough relationship $p = mc$ that we used in the discussion after (2.3), we have a crude estimate for the corresponding smallest mass-energy as $m_{\min} \approx 7 \times 10^{-67}$ g. This is the mass-energy of a particle-wave spanning the Universe. I am not aware of any very direct role of such wavelengths in current thinking, but perhaps they could be relevant to crudely modelling the large scale structure of the Universe as a quantum-mechanical system. Now in quantum mechanics, when one looks at it in more detail beyond the de Broglie formula, it often happens that the 'vacuum' or 'ground state', i.e. the state of a quantum system of lowest energy, does not in fact have zero energy. For a quantum oscillator, for example, the energy of the vacuum is half that given by the Planck formula (2.1) for a wave of the same frequency, and we might imagine something similar here. Dividing by the volume of the Universe we would have a vacuum mass-energy density of 2×10^{-152} g/cm^3 on the basis of a quantum oscillator of our maximum wavelength, or, equivalently, minimum mass-energy. In modern cosmology there does seem to be a mysterious energy-density Λ that needs to be added to the pure form of Einstein's equations relating matter and curvature. This is called the cosmological constant or

† We will crudely treat the Universe here as a box of some size whereas it is actually thought to be a 3-dimensional sphere, but doing it properly one has the same conclusion of a minimum momentum resolution. The associated momenta are non-commutative or 'quantum' as an expression of the curvature, and one can use the machinery given in Section 2.3 but applied to momentum rather than to spacetime. The precise sense in which this is an example of Born reciprocity is explained at the end of Section 2.4.

'dark energy', but unfortunately it is required at a very different value of around 10^{-29} g/cm^3.

Actually, our best understanding of particle physics is not quantum mechanics but quantum field theory in which, for each kind of particle in Nature, there are associated quantum oscillators of all frequencies, with a corresponding vacuum contribution from all of them. As a picture one can imagine that the vacuum is teeming with a 'sea' of virtual particles continually being created in particle–antiparticle pairs and recombining, forming quantum fluctuations in otherwise empty space and with a consequential vacuum energy density.† In this case it is not the smallest but the highest energy quantum oscillators (those with highest frequency or the shortest wavelength) which will dominate the net vacuum energy. Since l_{Planck} is the smallest meaningful length scale as we have explained above, we can take this as the shortest possible wavelength. Hence we have a corresponding vacuum mass-energy of around m_{Planck} or a density if spread over the Universe of around $m_{\mathrm{Planck}}/r^3 = 7 \times 10^{-91}$ g/cm^3 from such an oscillator, where r is the size of the Universe. The average mass-energy among different oscillators will be some fraction of this, but this need not concern us at the crude level of our calculation. And, how many such oscillators are there? The total 'volume' of the space of allowed spatial momentum in mass-energy units is m^3_{Planck} and since m_{min} is the smallest separation in momentum space in mass-energy terms as explained above, we have $(m_{\mathrm{Planck}}/m_{\mathrm{min}})^3 = (r/l_{\mathrm{Planck}})^3$ many oscillators, one for each cell of size m^3_{min} in the space of allowed momenta. Combining these results and cancelling r^3 between the expressions, the vacuum energy density for each kind of particle should be something like

$$\frac{m_{\mathrm{Planck}}}{r^3} \times \left(\frac{r}{l_{\mathrm{Planck}}}\right)^3 = \frac{m_{\mathrm{Planck}}}{l^3_{\mathrm{Planck}}} = 5.1 \times 10^{93}\ \mathrm{g/cm}^3.$$

This 'Planck density' could be interpreted as indicating an average of one Planck-energy oscillator per box of volume l^3_{Planck} in actual space,

† It is a similar effect that near the surface of a black hole causes Bekenstein–Hawking radiation.

but this is misleading since the quantum oscillators are localised in momentum and not in position. In terms of imagining a 'sea' of virtual particle fluctuations it is as if the most relevant ones have mass-energy m_{Planck} and typical separation l_{Planck}. It is the sort of answer one might naïvely expect for a quantum-gravity effect but we see that it errs now on the other side in being far too large by a vast factor of 10^{123}. It would appear that the observed value lies somewhere between our two vastly differing ideas. What we need is a theoretical rationale for something nearer to the observationally required value of the cosmological constant, which is close to $m_{\text{Planck}}/l_{\text{Planck}}r^2$. Put another way, we need a theoretical reason in our answer to replace the *smallest* area l^2_{Planck} of anything by the *largest* area r^2 of anything – it is hard to imagine being more badly wrong. The lack of any theoretical understanding of the required value is called the *problem of the cosmological constant*. Its understanding would be a good test of any putative theory of quantum gravity. If you like, we get here an estimate of just how big our hole in science is. Vacuum energy is also called 'zero point energy' and is a favourite of science fiction writers as a fanciful source of energy.

2.3 QUANTUM SPACETIME

The continuum assumption on space and time seems then to be the root of our problems in quantum gravity. It is tied up with the very idea of point particles, of being able to point to exact positions in space and time and others arbitrarily nearby. But has anyone ever seen an exactly *point* particle, I mean one of truly infinitesimal size? We have argued that the concept is in fact physically meaningless as separations below 10^{-33} cm make no sense. Surely these concepts were invented as a mathematical convenience or idealisation valid at everyday scales but as we have seen they are not appropriate as a fundamental structure for space and time. On the other hand, it is extremely hard to think of an alternative, some would say mind-boggling. If spacetime is to consist of 'foam' of size 10^{-33} cm, what are these bits of foam *in* if not a continuum? Or if spacetime is fuzzy due to quantum effects, what is it fuzzy with reference to if not a continuum of possibilities? The fact is that our everyday geometric intuition just is not up to it. How indeed can we have geometry without points in it? We must turn to mathematics to see how better to do geometry itself.

Mathematical interlude

The mathematics that we need is abstract algebra. Since this may scare the reader, let me say that the following kind of thinking is actually the historical origin of algebra† and is the first thing we learn in high school when we say 'let x be an unknown variable such that $x^2 = 1$' (for example) and then proceed to solve for x. What is happening here is that the symbol x is a placeholder for some possible number that we have to find, and which we do not yet know. When we write x^2 we are not yet multiplying actual numbers but the symbols x in an *algebra* generated by all possible sums and products of this symbol with itself and with ordinary numbers and with the same rules as would apply to numbers. We are using these rules in the algebra when we rearrange an equation to solve for x. Let us denote by $C[\mathbb{R}]$ the algebra made up of all functions in one variable x. This can be thought of as the algebra corresponding to the real number line, denoted $\mathbb{R}$ in mathematics. The algebra of functions in one variable x *with* the additional rule $x^2 = 1$ picks out the points $x = 1, -1$ if we solve it. So this second algebra is largely equivalent to speaking of the set of points $1, -1$.‡

Now let us do the same thing a bit more ambitiously, for the circle and for the sphere as shown in Figure 2.6. Every point P of the

† See the essay of Heller in this volume for a short account of the history.

‡ Some of my physics colleagues resistant to abstract algebra should note that they are happy to work with complex numbers. What is a complex number? It's just the same idea, we work with the same algebra of functions in one variable x but this time put the relation $x^2 = -1$. This time there are no solutions among ordinary numbers (the square of any nonzero real number is always positive), so this goes beyond ordinary geometry. But it's a perfectly good algebra which we can use in place of the imaginary number that does not exist. One simply gives x here a special name ι which now solves $\sqrt{-1}$. Personally, I would go further and note that all that really 'exists' are whole numbers. Then fractions and real numbers $\mathbb{R}$ are likewise invented structures built on these.

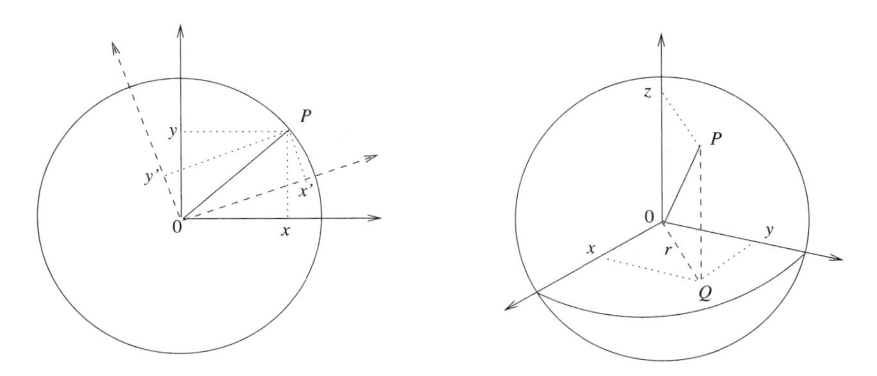

FIG. 2.6 Circle and sphere described algebraically in terms of coordinates.

circle is described by two numbers x, y which measure the distance you have to go along the x, y axes to reach P from the origin (ignore the other x', y' numbers in the figure for now). If we choose the numbers x, y arbitrarily we are talking about a general point on the plane. The circle is a particular subset of points, namely those with fixed distance 1 (say) from the origin. By the famous theorem of Pythagoras we know that the square of this distance, as the hypotenuse of the triangle, is $x^2 + y^2$, so the points of the circle are characterised by the equation $x^2 + y^2 = 1$. This was the warm-up. Now let us do the same thing for the sphere in the figure. This is described by three numbers x, y, z which measure the distance along the x, y, z axes from the origin. If we choose these numbers arbitrarily we are now talking about an arbitrary point in 3-dimensional space. The sphere is the particular subset of points with fixed distance 1 (say) from the origin. Now if you think about it, this distance is the hypotenuse of a triangle with vertical distance z and horizontal distance r, say, and the constraint is $r^2 + z^2 = 1$, by the theorem of Pythagoras applied to the triangle formed by 0, P, Q in the figure. But by the same theorem applied to the horizontal triangle in the figure, $r^2 = x^2 + y^2$. So the points of the sphere are characterised by the equation

$$x^2 + y^2 + z^2 = 1 \tag{2.6}$$

for any point (x, y, z) on its surface.

Now, let's turn all this around and think of x, y, z not as numbers but as 'unknown variables' or symbolic placeholders for actual numbers. They are the unknowns in the equation (2.6) if we try to solve it to find an actual point of the sphere. Once again, we take the algebra obtained by multiplying and adding symbols x, y, z in all possible ways but in the same manner as one would numbers, I mean subject to the same rules such as

$$(x+y)z = xz + yz, \quad x(yz) = (xy)z, \quad xy = yx$$

and so forth for any elements of our algebra (not just the examples shown). These rules are known respectively as rules of distributivity, associativity and *commutativity*. Just try out such rules for yourself with any actual numbers and you will see that they hold. Since x, y, z are symbols for some unspecified numbers, we impose on them the same properties. The algebra of such symbols subject to such rules alone is an algebra which I shall denote $C[\mathbb{R}^3]$ as they describe any point in all of 3-dimensional space corresponding to the three real variables. If we impose (2.6) as a further rule we have the algebra $C[S^2]$ corresponding to a 2-dimensional sphere S^2. In this way an actual space, here S^2, is replaced by an algebra of three variables with an additional relation (2.6). The same method applied to variables x, y and the equation for the circle gives the algebra $C[S^1]$. And just the same ideas, but this time hard to visualise unless you have 4-dimensional vision is $C[S^3]$ as the algebra for a 3-dimensional sphere. Its equation has the same form $x^2 + y^2 + z^2 + w^2 = 1$ in terms of four variables x, y, z, w. We see that the algebraic method is powerful and frees us of the need to have a picture in the first place. Assigning actual numbers to our variables picks out an actual point of the geometry in question *but we need never do this*. We can just work with the variables as abstract symbols.

For example, in high-school textbooks, you probably came across the formula for differentiation

$$\frac{\mathrm{d}f}{\mathrm{d}x} = \left((f(x+y) - f(x))y^{-1}\right)_{y=0}$$

where f is some function in one variable. One usually thinks about this geometrically in terms of the tangent to the graph of the function at an actual point x but this is totally unnecessary. One can just consider $\mathrm{d}/\mathrm{d}x$ as a certain operation that sends one element $f(x)$ in the abstract algebra $C[\mathbb{R}]$ (the algebra of functions in one variable x) to another. It is defined as above using at an intermediate stage the algebra $C[\mathbb{R}^2]$ of functions in two variables x, y. One can similarly define differentiation on $C[S^2]$ in such an algebraic way, it's just a little more complicated but the ideas are the same. The same applies to other geometric objects in terms of their corresponding algebras and to most geometrical constructions on them, including curvature.

Now, the great thing is that when you develop all this you discover that *you never really needed to assume the commutativity axiom*. If $xy \neq yx$ for any of the variables in the algebra, that means that these variables can never be set equal to actual numbers (since numbers always commute), so such an algebra does not have any points and goes beyond any geometry that you could ever *see*. But mathematically it still makes just as good sense.

To give a little example that some readers can have fun with, consider the above example of functions in one variable, but assume that $yx = qxy$ in the 2-variable algebra, where $q \neq 0$ is some number. Now take the very same definition as above and compute

$$\frac{\mathrm{d}x^2}{\mathrm{d}x} = ((x+y)^2 - x^2)y^{-1} = (xy + yx + y^2)y^{-1} = (1+q)x$$

on setting $y = 0$, instead of the answer from high school which was $2x$. If you have a bit more time, you can similarly show that

$$\frac{\mathrm{d}x^n}{\mathrm{d}x} = [n]_q x^{n-1}, \quad [n]_q = 1 + q + q^2 + \cdots + q^{n-1} = \frac{1-q^n}{1-q}. \qquad (2.7)$$

If $q = 1$ one has the same answer nx^{n-1} familiar from high school but this is the ordinary geometry case. When $q \neq 1$ the geometry here does not exist in the usual sense but we see that whatever this '*quantum differential calculus*' is, it q-deforms the classical case. In this example it is the 2-variable algebra needed to differentiate one variable

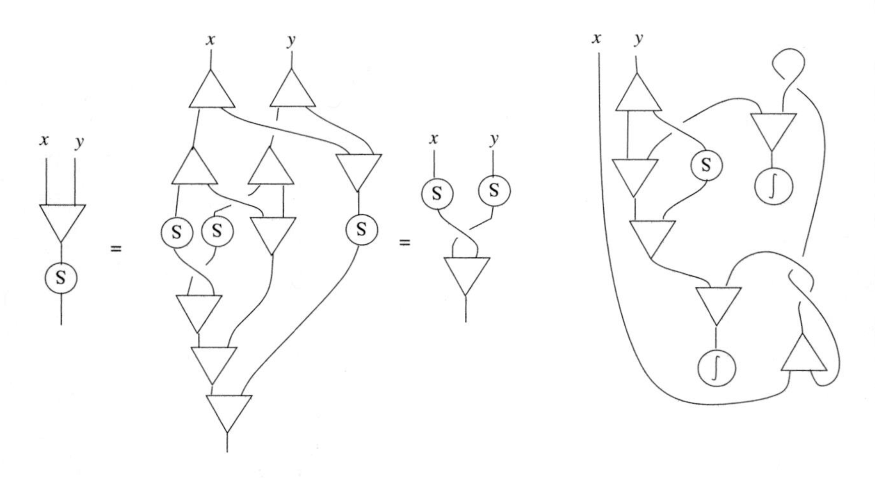

FIG. 2.7 A quantum version (left) of the proof in usual algebra that $(xy)^{-1} = ((y^{-1}x^{-1})(xy))(xy)^{-1} = y^{-1}x^{-1}$, from Majid (1995). On the right is a computation involving trace, two Fourier transforms, $\partial/\partial x$ and conjugation by y. As well as of relevance to quantum gravity, these could be circuits of an actual quantum computer in the twenty-second century. Unlike a usual computer, when one wire jumps over another there is a non-trivial (braiding) operation.

that is being made noncommutative. We can similarly write down the 'quantum sphere' $C_q[S^2]$ with three variables x, y, z as above and relations that q-deform the ones above in the sense that as $q \to 1$ one obtains $C[S^2]$. What is a remarkable fact, discovered in the late 1980s and early 1990s is that most of the basic geometric constructions in physics q-deform in a definitive and fairly coherent manner. Strictly speaking, the kind of algebra really needed here is one where independent variables do not necessarily commute either, as we have just seen in the example of differentiation above. It is what I called *braided algebra* or 'braided noncommutative geometry'† (the two terms are synonymous as the geometry is defined by the algebra) and it is best done for more complicated examples by means of 'braided flow charts' as illustrated in Figure 2.7. The idea is to think of the

† See Chapter 10 of Majid (1995) for a review of my research papers 1990–94 where this approach was introduced.

product of variables as a box with two strings coming in, one for each variable, and one string coming out for the result. You can then 'wire up' all algebraic expressions in braided geometry much as an engineer would wire up silicon chips in a computer, now as a 'flow chart' of the operations to be performed. The difference is that whenever a wire jumps over another wire, in braided geometry there is a non-trivial braiding operation, typically given by some large matrix depending on a parameter q. In the example above of the q-derivative there is a factor q whenever a wire goes 'over' another and q^{-1} when it goes 'under' (one has to fix a convention for which is which). Incidentally, these same methods have recently been proposed to actually build 'topological quantum computers' as a physical realisation of the same mathematics.

Now let us return to our classical examples for a bit more geometry. The problem with describing even an ordinary circle S^1 in terms of coordinates x, y in this way is that the actual numbers they stand for are somewhat arbitrary. The values depend on the precise angle at which we put the axes in Figure 2.6 – we can have another dashed pair of axes where the same point P is represented by new numbers x', y'. The true geometry of a circle refers to the points themselves and not to a meaningless pair of numbers. The way to get the actual geometry is to understand not only that we have chosen some orientation of the axes but to understand the rotational symmetry that relates different choices. If we both fix a system of axes *and* know how our expressions behave as we change them, we have all the information. Only expressions built from x, y that are unchanged as we change the axes are truly geometrical in classical geometry (for example the distance from the origin, which expresses the constant curvature of the circle in geometrical terms). In just the same way, the triple of numbers x, y, z describing a point of a sphere S^2 has no intrinsic meaning, it only has meaning in conjunction with the action of the group $SO(3)$ of 3-dimensional rotations that rigidly rotate one triple of axes to another. So any candidate for a deformed $C_q[S^2]$ would not be very convincing if there was not also some suitable object q-deforming

$SO(3)$. Such an object exists and is called the *quantum symmetry group* $C_q[SO(3)]$. The trick is that *there is no actual set of rotations of the q-sphere,* it is not an actual set of things and you could never actually see an individual q-rotation. What there is is a certain noncommutative algebra which if $q = 1$ would be the algebra $C[SO(3)]$ corresponding to $SO(3)$ as a space, and which has all the other properties, expressed algebraically, that you might want for rotations. It acts in the appropriate algebraic sense on $C_q[S^2]$.

Symmetries are perhaps the most important objects in all of mathematics. What we see is that any convincing 'quantum spacetime' also entails quantising this key notion. Whereas symmetries in classical mathematics are made up of individual transformations, for example the symmetries of an equilateral triangle are actual rotations and reflections of the triangle, a quantum symmetry refers to a quantum version of a whole collection or 'group' of symmetries and need not contain any individual transformations. Instead the collection as a whole is described using our algebraic techniques. M. Jimbo and the Fields medallist V. G. Drinfeld in the late 1980s essentially showed that all the 'basic' groups of symmetries that you can make smoothly, like the rotation group above, have natural q-deformed versions.† This has spurred a generation of pure mathematicians to study such objects for their deep and marvellous properties. The original physical motivation for them was from certain models of solid state physics rather than quantum gravity, but they can be applied to quantum gravity as we are about to see. I am not going to be able to say too much about the details of quantum symmetry groups other than to say that, first of all, they are algebras, typically noncommutative. Secondly, they have another structure Δ called the 'coproduct' which goes backwards from one copy of the algebra to two copies. So if the product is described by the joining of lines in a box ∇

† Their result was for all 'simple Lie groups' of symmetries, see Drinfeld (1987), and was actually for the infinitesimal rather than the geometrical picture discussed, which came later.

with two inputs and one output, the coproduct is described the other way by a box Δ with one input and *two outputs*. There are axioms expressed as diagrams which are entirely symmetrical between the two, a fact which we will later use to define a dual quantum group in which their roles are reversed. You can think (very sloppily) of the coproduct as being the set of things which, if composed as transformations, *would give* a given transformation. For example, Figure 2.7 actually shows some quantum group computations with both a product, denoted ∇, and coproduct, denoted Δ, and a braiding which you could take to be trivial. However, it turned out that all the standard q-deformation quantum groups have braided versions where exactly this kind of braided algebra applies with a non-trivial braiding when wires cross, and which then really brings out their deeper structure.[†] Aside from the q-deformation quantum groups and associated q-deformed and braided geometries, the other main class known so far are the bicrossproduct quantum groups introduced in Majid (1988b). Nowadays I tend to use a parameter λ to express this second deformation of classical symmetry groups, which are recovered as $\lambda \to 0$. Associated to them is a second λ-deformed 'flavour' of noncommutative geometries.

Finally, there are many other noncommutative algebras of all stripes, not necessarily deformations of anything and not necessarily involving quantum groups at all, but where one still has many new tools of 'noncommutative geometry'. You can see much more about one side of this and its application to physics in the essay of Connes. It is by now a large field and one can use the broader terms 'quantum algebra' or 'quantum geometry' to cover the many aspects of it.

Back to the story

That was our quick introduction to some of the mathematics, which has many applications within mathematics and elsewhere in physics.

[†] Such as a certain braided-self-duality relevant to quantum gravity in Section 2.5. See Majid (1995).

My proposed solution to our riddle of how to get rid of the continuum assumption then is to suppose that spacetime is not in fact a usual geometry with actual points but a noncommutative or 'quantum' one. It is because there are not actual points in such a 'quantum spacetime' that we have indeed let go of the continuum assumption. Moreover, pure mathematics is telling us that this is a very natural thing to do, we never really needed the points to do geometry, provided we do it algebraically as above. But there are also physical reasons to think that quantum spacetime is the right solution for matching up with the current scope of experimental possibilities today. In making this proposal I do not *necessarily* propose that physics be rebuilt on quantum spacetime as the ultimate theory of everything unifying quantum theory and gravity. Rather the way to think of it is as a new kind of geometry, one better than our everyday continuum notions and good enough to model *at least* the first quantum corrections to geometry that a full theory would predict. Personally, I believe it can tell us much more than that, but let us proceed cautiously for the moment.

The point is that whatever quantum gravity is, in one limit $l_{\text{Planck}} \to 0$ (more precisely when the relevant physical scale of study is large when compared to the value of l_{Planck}, we can't actually change its numerical value), the theory has to tend to ordinary geometry and gravity, the theory of curved spacetime and so forth. The first corrections to geometry should therefore be proportional to l_{Planck}, the next order corrections proportional to l^2_{Planck} and so on. But 10^{-33} cm is an incredibly small scale. If first-order effects are only now testable, it's unlikely that second-order ones ever will be without some completely new kind of experiment far from current thinking. So quantum spacetime is *at minimum* a pragmatic approach, it should suffice to lowest-order approximation and this is the most we can reasonably hope to test anyway at the moment. I would venture to say also that all currently envisaged experimental predictions of quantum gravity *should* factor through quantum spacetime and that we can understand a lot about quantum gravity at this pragmatic level without knowing the full theory. Indeed, because the difference between classical and

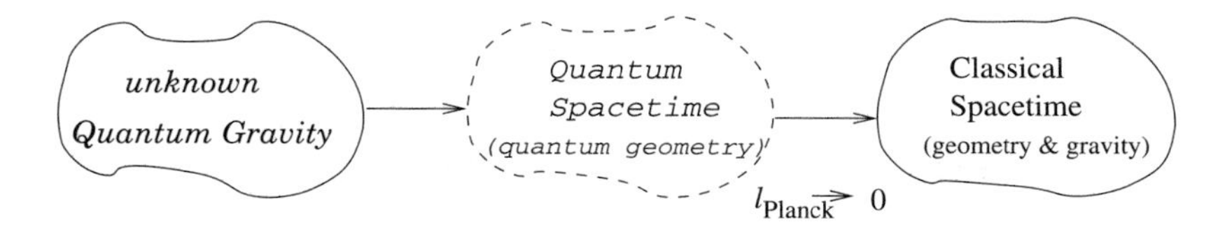

FIG. 2.8 Any theory of quantum gravity plausibly should pass through quantum spacetime on its way to ordinary classical spacetime. Hence quantum spacetime should be the framework for the first order in l_{Planck} experimental predictions whatever the actual theory.

quantum spacetime is exactly a noncommutativity of the form somewhat familiar in quantum mechanics (as we shall explain), and since the corrections are supposed to come from quantum effects messing up classical geometry, it is plausible that *any* quantum-gravity theory will indeed manifest itself at lowest order in this way:

> **Proposition:** Spacetime *is* noncommutative or 'quantum' as a new effect caused by quantum corrections to geometry. This should be an effective description one better than classical spacetime as this emerges from any theory of quantum gravity.

At the moment this assertion has the status of a logical prediction or argumented conjecture but if it could ever be confirmed experimentally it would be the discovery of a new physical effect every bit as great as the discovery of gravity. In fact for reasons to be explained in the next section, the noncommutativity of spacetime could be called 'cogravity'. It would not be true to say that the conjecture is entirely new at the level of casual speculation, but only in the 1990s with the discovery of quantum groups did the first actual models of possible quantum spacetimes appear. On the one hand we expect quantum gravity to reduce in the classical limit to ordinary classical spacetime, an actual geometry, and at the next order of accuracy to a noncommutative or quantum geometry. Conversely, if we do not know the full quantum-gravity theory, we can still study the possible 'quantum geometries' that could arise, which can be highly restrictive, as a guide to the unknown quantum-gravity theory. This is what I call the 'algebraic approach to quantum gravity' and which has been the topic of most of my own work in the last two decades.[†]

For the q-deformation type models as above, according to what we have said, we would expect something like

$$q - 1 \approx 2\pi (l_{\mathrm{Planck}}/r) \tag{2.8}$$

plus further terms with $(l_{\mathrm{Planck}}/r)^2$ and even higher powers on the right. This is because the deviation of q from 1 expresses the extent to which the spacetime is noncommutative or 'fuzzy' and we would expect this to be controlled by l_{Planck}. The latter is, however, a dimensionful quantity and what matters is not its absolute value but its value in comparison to the size of the object under study. Thus, r needs to be some other scale, the one of interest in the physical application. It is convenient to put in a factor 2π as we have done here to reflect the angular nature of q, where $\pi \approx 3.14$. At large scales $r \gg l_{\mathrm{Planck}}$ the spacetime will appear classical, while at Planck scales it will not. For example, if we are modelling the large-scale structure of the Universe we should take r to be the curvature length scale set by the cosmological constant or, which is about the same thing, the size of the Universe, at something like $r \approx 3 \times 10^{28}$cm. So in this application, using also our value (2.5) for l_{Planck}, we obtain something like

$$q - 1 \approx 3 \times 10^{-61}$$

which is a very very small deviation from 1. The actual amount of deviation is determined by the details of the model and can even be imaginary – the above should be viewed only as a first estimate based on dimensional analysis and the likely length scales at hand. In fact these ideas are only really tested in the case of quantum versions of 3-dimensional spacetime, where quantum gravity with cosmological constant is somewhat understood as a topological quantum field theory, a connection already remarked in the caption of Figure 2.7.

[†] See Majid (2006a,b) for recent reviews.

The relevant algebra for q-deformed models of 4-dimensional flat spacetime was developed by myself and others in the 1990s but it is quite complicated. Probably too complicated for experimenters and physicists to get their heads around in the first instance – one really needs the braided computation methods. Suffice it to say that a key feature is that now the four variables t, x, y, z for spacetime are such that t commutes with the others, but x, y, z do not commute among themselves, while d/dt behaves somewhat like in our example of q-deformed differentiation above.[†] Next, to capture the true geometry of flat spacetime, which is the main content of Einstein's theory of Special Relativity, means that we need to understand the Lorentz transformations $SO(1,3)$ that 'rotate' the axes in classical spacetime and the group of 'translations' that relocate the axes. The former is the origin of time dilation and Lorentz contraction when we switch from variables t, x, y, z to new ones t', x', y', z' much as depicted for the circle in Figure 2.6. We would also like to be able to relocate the origin of our axes and this would shift values, just as in the figure we would shift the values x, y if we relocated the point 0 where the axes cross. The translations and Lorentz transformations together form the *Poincaré group*, named in honour of the mathematician Henri Poincaré who obtained these results a little before Einstein. All of this goes through, there is a quantum group $C_q[SO(1,3)]$ which generates quantum time dilation etc., and a full (slightly extended) q-Poincaré quantum group built from these and a braided quantum group of q-translations.[‡]

Instead of saying more about the above model, I will now focus on the other family of deformations, where the quantum spacetime is

[†] See Chapter 10 of Majid (1995) for more details. It appears naturally as a braided q-deformation of 2×2-Hermitian matrices, a point of view inspired in part by twistor theory.

[‡] It turns out that the q-Poincaré quantum group here is forced to be slightly bigger and includes also *dilations* that magnify the variables t, x, y, z. Only massive particles in physics are sensitive to such scaling, it is not seen by light, for example, and is *fairly* harmless. It perhaps hints at a possible explanation for why the masses of particles are small compared to the Planck mass m_{Planck}. There is also a full q-conformal quantum group symmetry of which this is a remnant. Conformal invariance also plays a central role in Penrose's essay.

simple enough (not too noncommutative) that theoretical and experimental physicists have been able really to get stuck into what such a quantum spacetime predicts that could be tested. Since I first proposed it almost 15 years ago,† this 'bicrossproduct model' has attracted a fair amount of interest. In this model our x, y, z coordinates commute so the 'space' part of the quantum spacetime is actually undeformed. The time coordinate does not commute, so $tx \neq xt$ etc. We require instead

$$tx - xt = \imath\lambda x \qquad (2.9)$$

and similarly with y, z in place of x. Here λ is a parameter used to control the amount of deformation from classical geometry and $\imath = \sqrt{-1}$. In a quantum-gravity context a naïve estimate for its value might be the 'Planck time'

$$\lambda = \frac{l_{\text{Planck}}}{c} = 5.4 \times 10^{-44}\ \text{s}$$

(the units here are seconds). There is a corresponding λ-Poincaré quantum group so that all of Einstein's Special Relativity again passes over to this quantum spacetime and we shall say a little more about it later.‡ Its structure is that of a bicrossproduct, from which the model takes its name.

These relations are simple enough for me to want to show them to you and discuss them a little. The reason is that we still have to address the most important issue of all; spacetime might be some quantum space in this sense but how could we ever test this experimentally? One signature is the fact that x, y, z, t can never be actual numbers. In quantum mechanics there is similarly a noncommutativity of a position component x and the corresponding momentum component p which forces them both to become matrix operations (similarly for the other directions). One replaces the idea of x, p being

† Majid and Ruegg (1994).

‡ Some recent authors have been so impressed by this that they misleadingly use 'bicrossproduct model' and DSR 'deformed or doubly special relativity' interchangeably.

numbers by x, p being matrix operations or 'operators'† expressing the act of measurement in a certain sense (this is known as 'Heisenberg's form of quantum mechanics' or what he later called 'matrix mechanics'). All you need to know about matrices is that they are arrays of numbers that can be multiplied but as such need not commute. In particular, Heisenberg's formulation of quantum mechanics was based on the relations

$$xp - px = \imath\hbar \tag{2.10}$$

which when $\hbar \neq 0$ has no solution among ordinary numbers but does among matrices. Here the number $\hbar$ (pronounced 'h-bar') is related to Planck's constant introduced in Section 2.2 by $\hbar = h/2\pi$, and expresses the quantum nature of the algebra. These equations (2.10) imply that the act of measuring x, p can give different answers depending on the order in which you do it, and this translated directly into the famous 'Heisenberg's uncertainty relations' $\Delta_x\Delta_p \geq \hbar/2$ where Δ_x, Δ_p are the statistical variations in repeated measurements of the relevant quantities. The fundamental reason for the uncertainty is that matrices being a collection of numbers necessarily have a spread of numbers attached to them. In modern terms, Heisenberg's idea was that the variables in quantum mechanics form a noncommutative geometry. In that same sense now, we see that t, x have to be viewed as operators and the relations say that if we measure t and then x we get a different answer than if we measure them the other way around.‡ If Δ_x, Δ_t are statistical variations in repeated measurements of x, t

† One is often interested in infinite matrices and these are kept under control by thinking of them as operators acting on an infinite-dimensional 'Hilbert space'. This is explained further in the essay of Connes in this volume.

‡ I am reminded here of when I attended the play 'Copenhagen' by the leading British playwright Michael Frayn some years ago. While waiting for the curtain to go up I overheard a young man in the row behind, clearly not a physicist, trying to impress his girlfriend with an explanation of Heisenberg's uncertainty principle around which the play revolves. He explained that you can't measure where and when a particle is at the same time ... which is entirely wrong for Heisenberg but exactly what I had recently proposed with Ruegg. Suffice it to say that I was every bit as impressed as the girlfriend.

respectively, whatever that means, the analogy suggests something like a new 'spacetime uncertainty principle' $\Delta_x \Delta_t \geq \lambda |\langle x \rangle|/2$ for this model, where $\langle x \rangle$ denotes the average value. The same for the y, z variables in the role of x. Of course the error will be absolutely tiny as 10^{-43} seconds is a tiny time frame and it would be hard to see this directly. Moreover, the analogy with quantum mechanics does not really apply to spacetime itself since in quantum mechanics the time variable is not quantised, measurements are being made at some *point in time* whereas here t itself is subject to intrinsic uncertainties.

On the other hand, since one has differential calculus in the quantum case, one can still write down quantum waves in analogy with classical waves, and compute their propagation speed as a correction to the classical speed. The problem then is how to identify the quantum wave with its classical counterpart. It is an area which is still quite speculative, but in this model the quantum waves look very similar to their classical counterparts and one can identify them under the convention that t is always written to the left in any expression before identifying with its classical version. If we do it in this simplest way, we obtain the prediction that the quantum light wave speed depends a little on its frequency. *In the model above, higher frequency light travels a little more slowly.* For example, blue light travels just a tiny bit more slowly than red light. This VSL (variable speed of light) prediction comes out of quantum spacetime and I am not aware of any other actual theoretical mechanism that predicts it (of course one can randomly speculate on the possibility of VSL without having any coherent reason or mechanism for it). In our case we have argued that any theory of quantum gravity has to lowest order of correction *some* quantum spacetime, and it is a generic feature of such models that there will then be *some* correction to the speed of light. In the model above it takes a particularly simple form, but there are reasons to expect some correction in most cases.

Of course, the constancy of the speed of light is a central tenet of conventional quantum theory and gravity, so we see in this way one of the key things that quantum gravity requires us to give up.

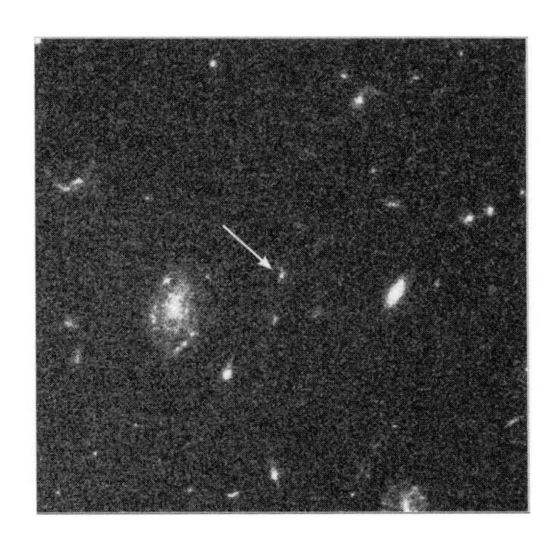

FIG. 2.9 Hubble telescope image with arrow to extremely distant host galaxy (about 12 billion light-years away) of gamma-ray burst GRB 971214 identified during a period of afterglow. The image is grainy because we are looking across to the other side of the Universe. Such bursts could provide evidence for quantum spacetime. Image credit S. R. Kulkarni and S. G. Djorgovski (Caltech), the Caltech GRB Team and NASA.

Such a bold claim should of course be tested. Now, without going into details, if we could make a burst of two types of gamma rays on the other side of the Universe, i.e. very far away, with two very different frequencies but as bright as possible, we could hope to see when the burst arrives at Earth. The higher frequency rays should arrive a little later according to what we have said. The reason it has to be on the other side of the Universe is that the effect is so tiny. The 5×10^{-44} s times a gamma-ray frequency of say 6×10^{22} vibrations per second gives a correction of 3 parts in 10^{21} to the speed of light. Only by multiplying this by a huge number, the 10^{10} years it might take to travel such a distance, can we hope to see a measurable delay in arrival time, in this case of about one millisecond (one thousandth of a second). Well, such gamma-ray bursts occur naturally. We do not know their exact origin, possibly the collapse of a binary system into a black hole, but we see them on Earth every day or two. They are extremely bright and short-lived; a gamma-ray burst can release more energy in 10 seconds than our Sun in its entire 10 billion year lifetime, and with a range of frequencies which can be up to the sort of level used in our estimate. Once it comes online, the NASA GLAST satellite will look for such gamma-ray bursts and part of its mission protocol is exactly to provide data for such 'time of flight' tests of VSL. There is just one catch, and it is a big one. Although extremely sharp by astronomical standards, the gamma rays are nevertheless not produced in a single instant and nor is their actual pattern of production understood with any accuracy. Therefore the arrival on Earth will in any case be spread out in a manner that we cannot control or accurately allow for. The tiny differences in arrival pattern expected in our prediction above are matters of milliseconds, so they would be very hard to see within that spread. The only thing to save us here is that the predicted effect is proportional to distance, so a statistical analysis of thousands of gamma-ray bursts looking for an effect correlated with distance could be used. This is quite a lot of work for astronomers. But in principle, by homing in on and identifying the host galaxy from which the burst originates (they usually have a visible afterglow that can be seen for example by the Hubble telescope) one can determine the distance from which it originates in order to make such an analysis.

Another experiment that could be done today, on Earth, to test for VSL, would be to adapt the technology currently in place for gravitational wave detection. These are very long baseline interferometers that split a laser beam into two, send one half around a long trip using mirrors and, when it returns, recombine it with the other half. A tiny variation in the structure of spacetime will cause the interference pattern made by the two beams to change. The same kind of experiment is being planned in space where an even bigger baseline will be possible (NASA's laser interferometry space antenna or LISA). The intention of these experiments is to look for ripples in the fabric of spacetime, but they could be adapted (at great expense) to test for quantum spacetime. The extreme sensitivity of the devices makes up for the much smaller travel time. What should be clear then is that the proposition that spacetime itself is quantum is logically expected, theoretically computable and in principle experimentally testable. Experiment and still deeper ideas from noncommutative geometry and quantum groups perhaps with no analogue in classical geometry, and a requirement that quantum theory and gravity redone on a quantum spacetime now be consistent could by themselves lead to a satisfactory

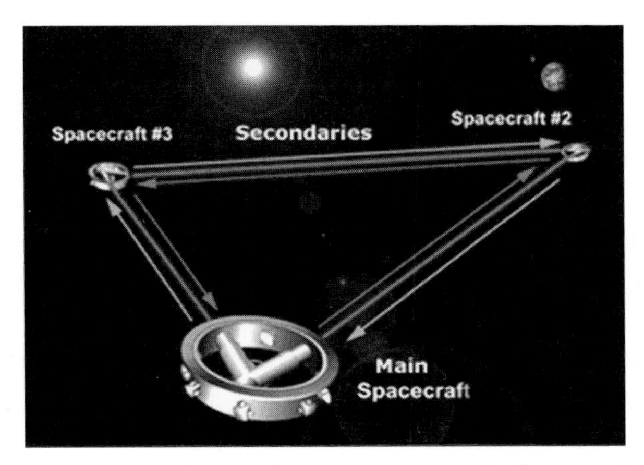

FIG. 2.10 Technology similar to NASA's LISA, which is intended to detect gravitational waves, could theoretically be adapted to test for quantum spacetime. Image credit NASA.

theory of quantum gravity, or they could remain steps towards a still deeper theory. It may be old-fashioned but, whether proved right or wrong, this is a healthy way to do theoretical physics in touch with all three of mathematical sensibilities, theoretical consistency and testability.

Finally, I do not want to leave the reader with the impression that quantum spacetimes need be merely deformations of ordinary ones. Much more incredible things are possible in this richer world, things with no classical analogue as hinted at above. I call them *purely quantum phenomena*.

The first one I want to mention is that the notion of differentiation in the quantum case can have some subtleties and one of them is that it may be necessary to have more quantum 'directions' than there are classically. For example $C_q[S^3]$ in the q-deformation family outlined above is a deformation of the algebraic description of a 3-dimensional sphere (it is almost the same as $C_q[SO(3)]$ already discussed). But if one wants the notion of differential calculus on it to be (quantum) rotationally invariant, one finds that it has to be 4-dimensional. So at each point on the quantum S^3 there are *four* directions in which you can move. It is a feature in fact of many sufficiently noncommutative quantum geometries and explored in Majid (2005), where I argue that the extra direction could be interpreted as *time*. In a sense, quantum geometries are dynamic not static objects that generate their own motion and it might be that we get here insight into *why things evolve in the first place*.† This is clearly speculative but in the spirit of this volume.

My second illustration, this time very specific to the q-deformation family is that the value of q where these quantum groups have links with quantum physics (in what is known as 'conformal field theory') is

$$q = e^{\frac{2\pi\iota}{n}}$$

where n is a whole number, $e \approx 2.71$ is a standard number coming out of differential calculus and $\iota = \sqrt{-1}$. What this means is that q is a number, different from 1, obeying $q^n = 1$. In this case the entire noncommutative geometry tends to have a reduced version where, if x is one of the variables in the geometry, one imposes further relations such as $x^n = 0$. Look carefully, for example, at (2.7) and you will see that $x^n = 0$ is compatible with such a formula for q-differentiation, but only if $q^n = 1$. If x was a real number, an equation such as $x^n = 0$ would imply $x = 0$ so in a sense imposing it leaves only the 'purely quantum' part of the geometry. These reduced 'purely quantum' geometries still have the flavour of their classical counterparts but are of finite size. For example, instead of all functions in x being allowed, now $1, x, \ldots, x^{n-1}$ form an independent set. In the same way, the reduced algebra, denoted $c_q[S^3]$, is a finite model of the 3-dimensional sphere S^3 and has size n^3 in the sense of that many independent degrees of freedom. This is the same size as the algebra of functions on a finite set of n^3 points. Whereas physicists, plagued with infinities in quantum gravity and elsewhere, often make ad-hoc lattice and other discrete approximations of spaces in order to make things finite, we achieve the same in a way that is fundamentally quantum and preserves the

† The modular group mentioned in Connes' essay has also been proposed as a kind of induced time, but it is not known if these two mechanisms are related.

structure of the geometry and any physics built on it.† Moreover, the real world might actually be like this with some value of q. Equation (2.8) in this context becomes

$$n \approx r/l_{\text{Planck}} \tag{2.11}$$

where r is the length scale of interest, which when modelling the whole Universe on a large scale (which at each instant in time is indeed thought to be a 3-dimensional sphere) is the size of the Universe, so $n \approx 2 \times 10^{61}$. Then the reduced quantum geometry has some 5×10^{183} degrees of freedom in it. We achieve here the effect of dividing the Universe into this many 'fundamental cells' of width l_{Planck} but without doing anything as ad hoc as that. At the end of Section 2.2 we crudely divided the range of energies of oscillators spread over the Universe into cells of mass-energy width m_{min} with one oscillator in each cell to again achieve such a finite number of degrees of freedom. The reduced quantum geometry achieves the same but more precisely and directly in the structure of spacetime. If we attribute a vacuum mass-energy density of 7×10^{-91} g/cm^3 to each degree of freedom as we did before, and multiply by our number of degrees of freedom, we get back to our previous incorrect estimate for the total vacuum energy, namely the Planck density 5×10^{93} g/cm^3. By contrast, taking n instead of n^3 for the true number of degrees of freedom to count here would give 10^{-29} g/cm^3, the observed value of the cosmological constant or 'dark energy' that we would like to explain. I do not have an argument for taking n in place of n^3 at the moment but perhaps one may emerge in time.‡ My general idea for the cosmological constant is that the correct theoretical explanation is that Einstein's equation without cosmological constant holds exactly but in a quantum spacetime using noncommutative geometry. When this is

† See Gomez and Majid (2002) where the theory of electromagnetism is completely solved on this reduced geometry when $n = 3, 5, 7$.

‡ There is a 'holography principle' suggested by string theorists in connection with black hole entropy, which says that the number of degrees of freedom in a geometry should be related to its bounding area, suitably understood, not its volume. This might suggest to take n^2 instead of n^3 whereas we would like to go even a bit further and take n.

expanded in terms of a correction to ordinary geometry, what appears is the naïve Planck density above but with a prefactor $(l_{\rm Planck}/r)^2$ (or $(1/n)^2$ in the above discussion) because, for reasons yet to be understood, this is a second-order quantum correction.

Where I think such 'reduced geometry' really ought to be used is on the surface of a black hole. We mentioned in Section 2.2 that in classical general relativity there is a theorem that black holes have 'no hair' in the sense of no imperfections or fluctuations of their surface. In that respect they are almost like elementary particles. But this is a classical theorem; there can be quantum states as part of any unknown theory of quantum gravity that reside on or very near the surface of the black hole. Even though we do not know the details of these states, one can use thermodynamic arguments to count them, in terms of their entropy S. Several bits of evidence point to a formula which I am going to write as

$$S = k4\pi(\frac{r_{\rm BH}}{l_{\rm Planck}})^2$$

where k is a conventional constant that need not concern us (Boltzmann's constant) and $r_{\rm BH}$ is half the radius of the black hole, related to its mass via the formula (2.4). Let us see how this might be understood using quantum spacetime. Because of the classical no-hair theorem the $C_q[S^2]$ quantising the black hole surface should have all of its 'classical part' factored out, so be modelled by the reduced geometry $c_q[S^2]$ with n^2 degrees of freedom. As before, we use (2.11) with the black-hole diameter $4r_{\rm BH}$ as the length scale r, so $n \approx 4r_{\rm BH}/l_{\rm Planck}$. Finally, without knowing much about these purely quantum degrees of freedom, we might suppose that they are approximately described by, say, a Gaussian probability distribution – as might be typical of a state in thermal equilibrium. Each such degree of freedom has entropy $k/2$,† which implies $S \approx \frac{k}{2}n^2$. This is close to the above expected

† The notion of entropy is discussed in Penrose's essay as a measure of volume in phase space. In our case it is important to view it statistically, so for each degree of freedom I have in mind that the integral of $-k\rho\log\rho$ is $k/2$ when ρ is any Gaussian probability distribution, irrespective of its details.

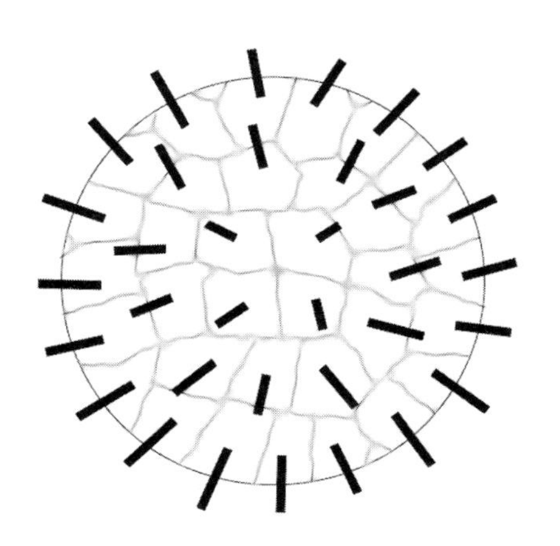

FIG. 2.11 Ad-hoc visualisation of the entropy of a black hole as due to 'quantum hair' with one degree of freedom per fundamental cell of area $l^2_{\rm Planck}$. This gives a flavour of why the entropy of a black hole is proportional to area but is ultimately meaningless since the cells are drawn and located *in* a continuum which is exactly what does not exist at this scale. Instead one may use a reduced quantum geometry to describe the surface of the black hole.

answer when we put in the value of n. I should stress that these are back-of-envelope estimates rather than actual derivations but they tell us that *black-hole entropy is plausibly consistent with an explanation as a reduced noncommutative geometry.* It makes rather more sense than dividing up the black-hole surface into a lattice of n^2 discrete points or Planck-area sized regions which would be a rather ad-hoc alternative (see Figure 2.11). Although the latter is easier to visualise, it only begs the question as to in what continuum space are these points or regions located.

2.4 OBSERVABLE-STATE DUALITY

Although quantum spacetime is a key prediction explained in this essay, I am not sure that we have yet found the *truly fundamental* answer to the structure of space and time. For one thing, we still do not know *which* quantum spacetime. For that we need a deeper understanding of quantum gravity and the equations that define how geometry and matter are interrelated. It seems likely then that we need some fundamental new insight to get us 'out of the hole' which, given that the hole has been there now for a good part of a century, needs to be something really really deep.

I have my own idea for this and it will occupy us for the rest of the essay. And when I say 'deep' I mean it has a fundamental or philosophical flavour which I will come to in Sections 2.5 and 2.6, while for now I am going to lead up to it in a more conventional way from within physics. From within physics one of the 'usual suspects' as to where we may have gone wrong is the nature of measurement in quantum mechanics, usually illustrated by means of 'Schrödinger's cat' and the Einstein–Podolsky–Rosen (EPR) experiment. I already mentioned in the preceding section that physical variables in quantum mechanics become 'measurement operations' in Heisenberg's formulation of quantum mechanics, which emphasises the particle aspect of a wave as having position and momentum (to the extent possible). The role of measurement is even more explicit in the equivalent 'Schrödinger's formulation' in terms of *wavefunctions*, which emphasises the wave aspect of quantum theory. These functions are spread over the range of possible values that a quantum measurement can result in and (roughly speaking) their amplitude at each point in the range encodes the probability of that result being obtained. Now, when the position of a particle is measured to obtain some actual number x_0, say, the wavefunction, which just prior to the measurement may have been spread out over a range of possible positions that might have been measured, is postulated in quantum mechanics to suddenly collapse to be sharply focussed at the point x_0. After all, if you were to measure the position an *instant* later you would surely get x_0 as you had just measured it. This framework is called the 'Copenhagen interpretation' of quantum mechanics as hammered out in a series of discussions between Bohr and Heisenberg in 1927, although it also owes a lot to Max Born who had argued for the probabilistic interpretation of Schrödinger's wavefunctions.

What is important is that somehow the act of a measurement has a physical impact. This is well-observed in experiments but it leaves one with huge philosophical problems as to where exactly to draw the line. What exactly is meant by a measurement? Does it matter if a machine makes the measurement but nobody looks at the answer, is that a measurement? It matters because the act of measurement is going to have a physical consequence. In 1935 Schrödinger put forward his famous 'Schrödinger's cat' thought-experiment to illustrate the difficulties in the form of a paradox. A cat in a sealed room gets a dose of poison gas if a certain atomic decay takes place and is registered in a detector in the room. The atomic system is described by a quantum wavefunction which gives the probability of decay at any given moment. After a while, there will be a good chance that the decay has taken place, so the cat may be dead already or it may still be alive. Viewed from inside the room the only quantum uncertainty concerns the atomic decay, but the cat itself is either dead or alive depending on whether the decay has happened or not at that point. But what if we don't look inside the room? We have to describe the atom, the detector and the cat by one joint wave function and so long as we do not look inside the room the cat's part of the wavefunction will still be spread between these two possibilities with some probability to be dead and some to be alive if one were to look in. Is the cat in this case dead or alive? The lesson I prefer to take from this is that quantum states do not have an absolute reality, they are a product of the use of quantum mechanics to model a situation and depend on the model you have in mind, in the same manner as the use of probability in statistics.

The EPR experiment was similarly put forward in 1935 as a putative thought-experiment intended as a paradox. A modern version could be to consider two photons created equally and oppositely at point *A*, in particular with opposite value of a quantum observable called helicity (this measures the direction of spin of the photon as it moves along). The two photons are detected at distant points *B*, *C* (in one experiment this has been done with *B*, *C* some 30 km apart connected to *A* by fibre-optic cable). At the instant that the detector at point *C* measures the helicity of one photon the wavefunction for the joint system must collapse so as to have the measured value of helicity at *C* which also entails having the opposite value at point *B* which, let us say, is about to be measured. So at the instant that the value at *C* has been measured we already know what value the detector at *B* is going to find. This appears to contradict the idea that information cannot travel faster than light. In fact, this effect does seem to be observed in actual experiments, but careful analysis shows that it is not possible to use it to transmit information instantly from C to B. But still, an instantaneous collapse of the wavefunction over an extended distance is worrying as it seems to defy the idea that physics is 'local' in the sense that there should be no spooky 'action at a distance'. I recall from Section 2.2 that this was likewise an objection to Newtonian gravity and led to the curvature of spacetime as its replacement.

These issues left the early pioneers of quantum theory including Bohr and Schrödinger to argue that a new (more Eastern) world view was needed to comprehend what was actually going on in quantum mechanics, that quite simply the quantum world has a different notion of reality. Heisenberg wrote already in one of his early papers:†

> The existing scientific concepts cover always only a very limited part of reality, and the other part that has not yet been understood is infinite. Whenever we proceed from the known into the unknown we may hope to understand, but we may have to learn at the same time a new meaning of the word 'understanding'.

Similarly, he later countered Einstein's famous criticism of the Copenhagen interpretation with an explicit attack on scientific realism or, as he put it, the 'ontology of materialism' implicit in Einstein's argument. At root here is the conviction that somehow the observer needs to be a partner in the very concept of reality. Yet this vision has not really come to pass; the measurement postulate is typically assumed by working scientists to be an approximation of a less mystical theory of measurement the details of which could be found in principle. After all, quantum mechanics is only a nonrelativistic special case of much more sophisticated quantum field theory. As a result most scientists do still assume today that there is a fundamental

† Heisenberg (1927).

observer-independent physical reality 'out there' and that we do experiments and make theories to come closer to it, a view going back to Bacon, Hooke and to the historical concept in the West of what is science. My view is that for quantum gravity, however, the measurement problem is more serious and can't be brushed off. This is because the division into atomic system and observer is also a division into micro and macro, which quantum gravity has to unify. If so, then perhaps we can't get further without understanding the realist-reductionist assumptions made in science more deeply. Perhaps our naïve or confused adoption of them is just plain wrong and has boxed us in as we saw in Figure 2.5. These warnings provide a clue to which we will come back later. In fact I shall not argue that we abandon our realist-reductionist assumptions but only that their deep structure feeds into the structure of quantum gravity.

Interestingly, the need for the 'observer to get into the act' was *also* key in Einstein's formulation of gravity, as well as for his earlier theory of Special Relativity. Let us start with the latter as a warm-up. It is well known that a charged particle moving in a magnetic field experiences a force perpendicular to its line of motion. This is why an electric current in an electric motor makes it turn. However, if you were moving along with an electron, from your perspective the electron is not moving, so there is no such force. Is there a force or not? The answer is that there is, but in the second perspective what was previously seen as a magnetic field will appear in part as an electric field and it is this part which causes the perpendicular force. This means that electric and magnetic fields are not absolute things any more than motion is an absolute thing; *you* the observer have to say what frame of reference you are doing your calculations in and in that frame of reference you will see one bit of physics, in another you might see something else. What exists is the pair consisting of the electric and magnetic fields together, in the same way as the components x, y, z for the position of a point on a sphere in the previous section depended on the location of the axes so have no intrinsic 'reality', but do describe an actual point once the axes are fixed. What is 'real' somehow exists in the cohesion between the different points of views from all observers – it is the true geometry of the situation. The mathematician Riemann in 1865 took this much further with a formulation of geometry in which not just rotated axes but any somewhat arbitrary parametrisation of the points of a space were allowed. In spacetime terms not only observers moving with different velocities should be allowed but any observers at all, even accelerating ones. This led to Einstein's formulation of gravity in his theory of General Relativity. In this theory the reality observed by different observers can be very different, which is a similar story to the quantum-mechanical conclusion but for gravity. For example, a person falling into a black hole appears to a distant observer to slow down and hover just above the black hole, never actually falling in. But from the point of view of the person, he or she does fall in and get crushed in the singularity in the centre of the black hole in a finite time. So does the person die or not? Even more worrying, for a very large black hole the point of no return is crossed without any particular warning; physics appears to the infalling observer fairly normal at this point with only mild gravitational fields. As with EPR one does not get a causal contradiction but it is unsettling. But this time, unlike for Schrödinger's cat, we have a language – geometry – that fully resolves the problem. There are different local coordinate systems for the geometry and one of them happens to be incomplete, but this does not change the reality of what happens provided one thinks in a coordinate-free and atemporal way about it.[†]

The full story here, which goes back to the cleric and idealist philosopher Bishop George Berkeley commenting in 1712 on an experiment of Newton, was taken up by the late-nineteenth-century positivist philosopher Ernst Mach and then influenced Einstein. Would the concavity of the surface of water in a rotating bucket of water still be there if there were no 'fixed stars' up in the sky? It is a strange

[†] My proposal for quantum theory is that one similarly needs a geometric language to handle 'change of coordinates' between one observational framework and another. Quantum spacetime which is both geometrical and quantum gives us a hint as to how this could be done.

FIG. 2.12 George Berkeley 1686–1753 (left) and Ernst Mach 1838–1916 (right).

question but Newton's explanation, which is also the usual high-school explanation is that the rotating water experiences centripetal forces needed to maintain the circular motion and these push up the water more where it lies nearer the rim. But how do you know the bucket is rotating? If there were no other particles in the Universe, what would the bucket be rotating with respect to? Berkeley and later Mach argued that it is in fact the existence of all the other matter in the Universe that must therefore be responsible for this effect. In a sense they turned out to be right: in our modern understanding of gravity there is no such thing as absolute motion. An equally good explanation from a modern point of view is that the bucket is not (from the point of view of an observer that rotates with the bucket) rotating, but instead the stars in the sky are moving around the bucket for this observer. Their collective effects presumably conspire to result in stationary water that is still pushed up in the observed way. This is a manufactured example and I'm not sure that anyone has ever really done the calculation in General Relativity in this way but a similar one is the Lense–Thirring effect which *is* well studied. Here a rotating spherical shell of mass induces Coriolis forces inside the shell which is not quite the same thing but is a step in the right direction. For our purposes it illustrates how things far away could be viewed as conspiring to create a bit of perfectly ordinary physical reality. And the

Lense–Thirring effect is not the act of one bit of the rotating shell but a collective act known as solving a certain differential equation over all of spacetime. This will be another clue to which we shall return later, that collective action can translate into ordinary physical perceptions.

The differential equation here is Einstein's equation[†] whereby the matter (more precisely, energy-momentum) distribution throughout the Universe determines the amount of curvature and hence, after solving, the geometric structure of any given region of spacetime. Other particles then move in the resulting spacetime (while themselves contributing to its curvature) so in a sense the notion of spacetime could be taken out of the equation as an intermediary concept as Mach might have wanted. Einstein wrote of this[‡]

> It is contrary to the mode of thinking in science to conceive of a thing (the spacetime continuum) which acts upon itself but which cannot be acted upon ... in this way the series of causes of mechanical phenomenon was [in Mach's ideas] closed, in contrast to the mechanics of Newton and Galileo.

This quote brings us to another of Mach's ideas, inspired by Newton's third law of mechanics, that every action should have an equal and opposite *back-reaction*. This in its wider application is sometimes called 'Mach's principle'. In the case of a particle, if its acceleration can only be detected with respect to another particle, a Machian argument could be that somehow the presence of particle 2 causes the acceleration of particle 1. But reversing their roles one will have an equal and opposite acceleration of particle 2 caused by particle 1. This will be our third clue, that the action and reaction of mechanics could have their origins in a kind of symmetry between observer and observed.

Returning now to quantum gravity, since the concepts of observer and observed are so fundamental to both quantum theory

[†] Covered in detail in Penrose's essay.

[‡] Einstein (1950), page 54.

and gravity, they should have a fundamental role in quantum gravity. About twenty years ago[†] I put forward some simplified models of quantum theory combined with gravity, precisely in order to explore this idea. In these models we restrict attention to curved spaces with a lot of homogeneity – 'homogeneous spaces', as would be appropriate if we are interested, for example, in the large-scale structure of the Universe. And we restrict ourselves to the quantum mechanics of a single particle on such a space. We recall that a single particle is described by a 6-dimensional 'classical phase space' consisting of 3 dimensions for the position of the particle and 3 more dimensions for its momentum vector (which is closely related to its velocity). We have explained in Section 2.3 that the Heisenberg algebra (2.10) quantises these variables to define the 'quantum algebra of observables' or *quantum phase space* in the case of flat spacetime. We now need the analogue of this for homogeneous curved spaces, i.e. in the presence of a bit of gravity. Such spaces have by definition a lot of symmetry as expressed in the existence of a 'momentum group' G, the action of which generates uniform motions on the position space M. This is the physical meaning of the momentum variable that we discussed previously, as generating motion in position space, and actually this is more precisely all I mean; a homogeneous system (G, M). The quantum phase space for a particle moving in such a position space M with momentum group G is a standard algebra which I will denote by $H(G, M)$ or simply H for brevity. The role of position variables is now played by the variables used in an algebraic description of M in the manner explained in Section 2.3, while the role of momentum variables is played by the infinitesimal generators of the group G. Finally, there are 'commutation relations' between position and momentum variables analogous to (2.10).

What I showed in this context is that there is a class of such homogeneous systems with the property that the quantum phase space H for the particle is a quantum group. As explained in

[†] Majid (1988b).

Section 2.3, what this means is that there is a map Δ going from one copy to two copies of H. Now think about quantum measurement. A modern way to do this that avoids getting bogged down with wavefunctions is to do things directly in terms of the average or 'expected' values that a quantum state entails. In a given quantum state every observable a in H has some average value $\langle a \rangle$ in the notation used before. But let us think of this assignment of the average value to every observable as a map from $H \to \mathbb{C}$ (where $\mathbb{C}$ means that we allow $\iota = \sqrt{-1}$ into our number system for the values that the map can take). This is what mathematicians (as opposed to physicists) typically mean by a quantum state and is somewhat equivalent to a statistical ensemble of wavefunctions. Now, what Δ means physically is that given two such maps ϕ, ψ we can build a new one $\phi\psi$ as follows. Its value at any observable a is given by first applying Δ to a. The result lives in two copies of the algebra H and we then apply ϕ to one copy, ψ to the other and multiply the results together.[†]

So, the quantisation of these homogeneous systems (G, M) has the feature that the set of quantum states on the quantum algebra themselves generate an algebra. This turns out to be the quantum algebra of *some other* homogeneous system. So if a quantum particle in our original space M is in some state ϕ, in which according to quantum mechanics one might say that an observable a (say of position or momentum) has the average or expected value

$$\phi(a),$$

another person could equally well say that ϕ is an observable for a particle in some other quantum system, that a is in fact a quantum state for this other system, and that the same numbers above should be written instead as

$$a(\phi).$$

[†] This product is a quantum version of the statistical convolution of probability distributions on a classical group.

I've glossed over a technicality here: states and observables in quantum theory have technical positivity and self-adjointness requirements, but this just means that in reversing roles in this way we may go, for example, from a state to a combination of observables. But what it tells us is that in these models there is no true 'observer' and 'observed' in the sense that the same numbers, which I regard as the only hard facts on the ground, can be interpreted either way. Thus,†

> **Proposition:** In the context of quantum phase spaces which happen to be quantum groups, there is possible a dual interpretation in which the roles of observables and states are reversed. The existence of this possibility is a compatibility constraint between quantum theory and gravity in so far as expressed in the curvature of phase space. The roles of position and momentum in the dual model are also reversed.

Indeed, for our homogeneous quantum phase spaces $H(G, M)$ to be quantum groups, one finds that firstly, the position space M has to be itself a group and has to 'back-react' on G. Then the dual homogeneous system has the same form as above but with G now in the role of position space and M in the role of momentum group generating motions in it. The dual system (M, G) then has quantum algebra of observables $H(M, G) = H(G, M)^*$ in the notation above, i.e., with the role of the two groups interchanged. Moreover, to have a quantum group the action and its back-reaction have to be be compatible in the sense that one determines an equation that the other has to solve. This is all reminiscent of Einstein's equation where matter tells spacetime how to curve and spacetime tells matter how to move, as well as clearly Machian. Our action and back-reaction obey a pair of cross-coupled differential equations which, while not being exactly Einstein's equation, have some similar features such as the solutions being typically forced to have singularities when the regions in which

† Majid (1988a).

the equations are to hold are unbounded. I consider them a baby version in our quantum groups paradigm of Einstein's equations. This is a microcosm of the full problem of unifying quantum theory and gravity, but already illustrates a deep principle to be developed further in the next section.

I would like to show you only the very simplest example in this class of models. Here M and G are both ordinary real numbers $\mathbb{R}$ with the law of addition, so their corresponding algebras have one variable x for position and one variable p for momentum. Instead of the Heisenberg commutation relations (2.10) we have relations of the form

$$px - xp = \imath\hbar(1 - e^{-\frac{x}{r}})$$

and the algebra H with these relations, which is the quantum algebra of observables, is a quantum group with a certain coproduct Δ. Here r is a certain length parameter to be determined. At large $x \gg r$ the algebra becomes (2.10) or ordinary quantum mechanics because the exponential of a large negative number is very small (just as 10^{-6} means one millionth). But at small x the algebra is different and this modifies how a freely falling particle with momentum p would behave. In particular, one finds that *it takes an infinite amount of time for a particle moving in from large x to reach* $x = 0$. Looking at the motion in detail there is a similarity with the formula for a particle falling into a black hole, where it also takes an infinite amount of time (as observed by an observer far away) to fall in. We can say roughly speaking that the model has some of the features of a black hole of mass M and size r related by the formula (2.4). Putting in this formula as well as the formula (2.2) for the Compton wavelength of a particle of mass m allows us to quantify these approximations. The result can be summarised by

$$mM \ll m^2_{\text{Planck}} \longrightarrow \text{usual flat space quantum mechanics algebra}$$

$$H$$

$$mM \gg m^2_{\text{Planck}} \longrightarrow \text{classical algebra of a curved phase space}$$

where m_{Planck} is the Planck mass of the order of 10^{-5} grams. In the first limit the particle motion is not detectably different from usual flat space quantum mechanics beyond the distance of one Compton wavelength from the origin. In the second limit the estimate is such that quantum noncommutativity would not show up at scales larger than the one set by the curvature. Here we would be seeing ordinary variables x, p for position and momentum locations (their combined range of values is the phase space) but as we have explained there is still classical gravity-like geometry in the model and it appears here as curvature. This is why I call H here the 'Planck scale quantum group'. The dual H^* has just the same form but with the interpretations of x, p interchanged and new values of the parameters in the dual theory in terms of the values above. The quantum algebra of this dual system is equivalent to Δ, i.e. to the curved geometry for our original system. The observable-state duality says that the quantum states of our original system are built from the quantum variables of the dual system and vice versa.

But there is more. Not only is H a quantum group, *the mathematical requirement to form a quantum group* extending (in a precise sense) the algebra of functions in one variable x for position by another variable p for momentum, *necessarily has the two-parameter form above,* for some parameters $\hbar, r$ (and which we have then carefully interpreted as above). In other words, this is everything that is mathematically possible among structures in a certain paradigm with observer–observed symmetry. So this philosophical structural requirement alone implies both quantisation and gravity as expressed in the parameters $\hbar, r$ being thrown up in the analysis of the mathematical possibilities. Roughly speaking, the quantum theory of our original system corresponds to the curved geometry ('gravity' in some rough sense) of the observer–observed reversed system, and vice versa. You can't have one without the other and maintain a self-dual form, so this is a feature of quantum gravity not visible in quantum theory or gravity alone. Moreover, in the light of our previous discussion we see, very roughly speaking, that gravity measures quantisation of the matter and vice versa the motion of the quantised matter measures the geometry of the background gravity. *In summary, observable-state duality offers a deeper principle from which quantum theory, gravity and perhaps Einstein's equation might be derived.*

I mentioned in Section 2.3 that there are at the moment two main families of quantum groups, the q-deformation ones and the bicrossproduct or λ-deformation ones. The above is how the bicrossproduct ones were discovered, so even if one disagreed with all of the philosophy and physics, thinking of the problem of constructing quantum groups as a toy version of quantum gravity was useful. Prior to the mid 1980s the axioms of a quantum group or 'Hopf algebra' had been written down, as far back as 1947 by the topologist H. Hopf, but had remained something of a curiosity or technical tool due to lack of a significant class of true examples. Moreover, the quantum spacetime (2.9) of the preceding Section 2.3 has, as mentioned, a quantum version of Einstein's Special Relativity expressed by a quantum group built from deformed translations in spacetime and Lorentz rotations $SO(1,3)$. It has the form $H(M, SO(1,3))$ where M is a certain 4-dimensional group. One could think of the quantum group therefore as the quantisation of a particle moving in M under the action of $SO(1,3)$. Moreover, M back-reacts on $SO(1,3)$ and this is essential to its construction and hence to the quantum version of Special Relativity. In short, for quantum spacetime the Poincaré symmetry quantum group has itself to be some kind of quantum system and the above makes this precise. Both the spacetime *and* its symmetries are quantum systems in this way.

There is one more comment I would like to make about quantum groups in this particular context. A criticism we had in Section 2.2 was that conventional physics assumed a continuum space or spacetime to begin with. We have been no less conservative in our toy models above – we considered the quantisation of particles moving on a classical (homogeneous) space M as our starting point. However, there is no special reason to do this. The pair of cross-coupled differential equations between the action and back-reaction of the position and momentum groups on each other (which was our toy model of Einstein's equation) turns out to be equivalent to a local factorisation of a larger group, which I shall write as $X = G \bowtie M$. I think this is a deep point of view: when something factorises into pieces then there is generally a sense in which each factor acts on the other as follows. Consider m in M and g in G multiplied in X in the *wrong order*, then $mg = g'm'$ for some new elements which we consider as g transformed by m and m transformed by g respectively. This is a kind of mechanism for Mach's principle – a factorisation induces an action and a back-reaction between the factors. I had once hoped to obtain exactly Einstein's equation (not some baby version) as some such factorisation (of the right kind of structure) but the best I can say is that many nonlinear integrable systems of differential equations *can* be cast in terms of a certain factorisation of infinite-dimensional groups[†] and this applies in principle to gravity in *three* dimensions or a part of gravity in four dimensions, so it is not the whole story but a good part of it.[‡] Moreover, with this formulation, all of the above can be done with G, M themselves quantum groups to begin with. So if $X = G \bowtie M$ where *now* X, G, M are quantum groups then we have a new quantum group $H = M^* \blacktriangleright\!\!\triangleleft G$ analogous to H above, where M^* is the dual quantum group to M in which the roles of product and coproduct are interchanged. The result has the observable-state duality in the form $H^* = M \triangleright\!\!\blacktriangleleft G^*$ where G^* is similarly the dual of G. These constructions are also from Majid (1988b) and H is called the *semidualisation* of X. We will return to it in looking at quantum gravity in the next section, albeit again in three spacetime dimensions.

Finally, we have been talking about 'quantum phase space' because nobody should dispute noncommutative geometry in this setting – Heisenberg's algebra with relations (2.10) is already noncommutative. We have seen above how gravity-like features were encoded

[†] Riemann–Hilbert factorisation.

[‡] One can also look at so-called sigma-models or (very roughly speaking) strings moving in a group $G \bowtie M$ and in this case the duality above can be interpreted as Poisson–Lie T-duality, which is also a kind of micro-macro duality, see Beggs and Majid (2001).

in Δ as curvature of phase space. I want to explain now that this is a general feature and applies equally well to the quantum spacetimes that we studied in Section 2.2.3. The first thing to say is that when a continuous symmetry group like $SO(3)$ is regarded as a space in its own right, it is curved. This can be attributed to the fact that the group law is itself noncommutative (one says nonAbelian in this context). Try rotating something about one axis and then about a different axis. The result is different if you perform the rotations in the reverse order. Another example is the 3-dimensional sphere S^3, which is manifestly curved, and can be identified with a certain nonAbelian group. In this way such groups provided historically the simplest examples of curved spaces. In the same way then, $C_q[SO(3)]$ and $C_q[S^3]$ are quantum groups and their coproduct Δ encodes the fact that they are the simplest examples of *quantum curved spaces*. We see that quantum groups are indeed quite generally a microcosm for the unification of quantum ideas and the idea of curved spaces. When the space in question is or is part of spacetime then the curvature here refers precisely to gravity. Now let us combine this with the idea of observable-state symmetry expressed as quantum group duality:

> **Proposition:** In the context of quantum spacetimes that happen to be quantum groups, quantum group duality interchanges the role of 'quantum' and 'gravity'. Quantisation of spacetime and gravity in this context are observable-state dual to each other.

For example, the bicrossproduct quantum spacetime (2.9) was flat (but quantum), hence its dual is a certain classical (but curved) space. Physically it is the momentum space for the model and indeed acts on the quantum spacetime with relations generalising Heisenberg's relations (2.10). This is a general feature: *observable-state duality applied to quantum spacetimes which happen to be quantum groups is a quantum version of Born reciprocity interchanging the roles of position and momentum.* In the bicrossproduct case the model is flat and there is no actual gravity, instead the curvature is in momentum space. Thus, quantisation of spacetime in simple models like this has a very simple interpretation as what I call *cogravity* – it means 'gravity' in momentum space.

2.5 SELF-DUAL STRUCTURES AND PLATO'S CAVE

Douglas Adams' radio script *The Hitchhiker's Guide to the Galaxy* is a favourite among many theoretical physicists of my generation, certainly a personal favourite of mine. Part of the allure is spot-on comic satire aimed at theoretical physicists and philosophers, reminding us to take ourselves not *quite* so seriously. When a physicist proclaims that they are 'seeing into the mind of God' they are taking themselves too seriously. On the other hand, if we want to know what the *ultimate* nature of space and time is, we are in fact after insight into the ultimate nature of physical reality and we should not be afraid to discuss it even if our ideas are tentative and not something we would publish in our regular research. Adams presents the answer to the ultimate question of life, the Universe and everything as, famously, *42*. I have already provided my concrete answer in the preceding section but now I want to explain the question properly at a philosophical level.

Let me note first that while we can proceed as we have done in previous sections building up incrementally towards a theory of quantum gravity and other forces, such a normal 'bottom-up' approach may take some decades and even if we succeeded by modern standards, a few centuries from now it would likely be outmoded. Surprisingly, I think we can jump ahead and get a glimpse of a somewhat *final* answer without all the bits in between. My strategy rests on something that pure mathematicians tend to know but which physicists in my experience do not: *it is often easier to prove an assertion in general than for a specific example*. When I was based in Cambridge I recall a respected senior physicist coming into the tea room, banging his fist on the table and proclaiming in my direction 'if it's got a theorem, it's not physics' and walking out. This was friendly banter but it missed the point. The reason that it's often easier to prove a general theorem than a specific example is that one need only make use of the axioms of the very things being studied. The more general the theorem, the fewer axioms and premises you have to deal with and the closer the proof is to a tautology. Therefore, when faced with a tough problem I always suggest to look at it from 'both ends', from the specific problem on the one hand and from general arguments on the other. So in our case what I want to do is a radical 'top-down' approach based as it were on stepping away and thinking more objectively about *what the mathematical structure of the ultimate theory of physics should be*, starting from the most general principles.

Indeed, physics takes place within mathematics; this is essential for the logic and accuracy needed for the 'hard sciences'. But Nature does not necessarily use the maths already in maths books, hence theoretical physicists should be prepared to explore the tableau of all of pure mathematics and pick out objectively what is needed, not merely the bits of structures they stumble upon by accident or fashion (which is what tends to happen). This requires a very different mind set from the usual one in physics. It is true that theoretical physicists can eventually take on new mathematical structures, for example quantum groups. But even as this new machinery becomes 'absorbed' the mind set in theoretical physics is to seek to apply it and not to understand the conceptual issues and freedoms. By contrast a pure mathematician is sensitive to deeper structural issues and to the subtle interplay between definition and fact. It is precisely this subtle interplay which will be at the core of this section.†

So in the tableau of all mathematics is there some subset that is physics? If so, what is the 'meta-equation' or 'principle' that characterises it? If we knew this meta-equation, we would have the mathematical structure of the ultimate theory of physics and 'all' that would remain would be putting colourful names like 'electron' and

† I am arguing here that for quantum gravity one really needs a 'Department of *Pure* Mathematics and Theoretical Physics'. It's a department that would have been more at home in earlier eras when mathematics and natural philosophy were as one (go back to medieval times for example). Their separation into often warring factions during the twentieth century is in my view part of the reason for the 'hole in science' being open for so long.

'proton' to the various bits of mathematics. Now my idea for this meta-equation is that as the 'ultimate equation' it must express the quintessence of what physics ultimately is. But what physics, as it is conceived today, ultimately *is* is the product of a certain way of investigating the world. This is the work of many minds over the centuries but a good part of what is Western science was, for example, consciously formulated in the seventeenth-century Scientific Revolution. It is not for me to make a historical case here, I just want to say that when Bacon, Hooke, Newton and others swept aside religion as the traditional grounding for what was 'true' (as it had largely been at the time) they did not replace it by nothing, they replaced it by another set of terms of reference and rules of engagement that had
[illegible] s in the works of Latin, Greek and, later on, Islamic
[illegible] t necessarily being aware of it, we inherit them today
[illegible] ciences as an intrinsic part of what we do. I am *not*
[illegible] rules of engagement or basic requirements of a phys-
[illegible] ere fabrications of human beings (I am not going to
[illegible] alist position) – I think they are unavoidable and 'out
[illegible] e way as pure mathematics is probably 'out there' waiting to be discovered (more about this in the next section). *But* this does not stop us seeking to be more self-aware of these rules of engagement or basic requirements about the logical structure of any theory of physics. If we do it right, this should be our meta-equation as the *sine qua non* of physical reality. I then see two possibilities at this point. The first (which I shall argue for the sake of simplicity) is that all that remains under something like this meta-equation *is* physics – then we will have proven that the ultimate theory of physics is to first approximation no more and no less than the equation of looking at the world in a certain way called being a physicist. This does imply a certain new outlook on the nature of reality in line with our preliminary 'clues' explained at the start of Section 2.4. The other alternative is that there are further laws or constraints of physics that Nature knows about that cannot be explained in this way. In that case what follows in this section at least cuts down what we have to search for,

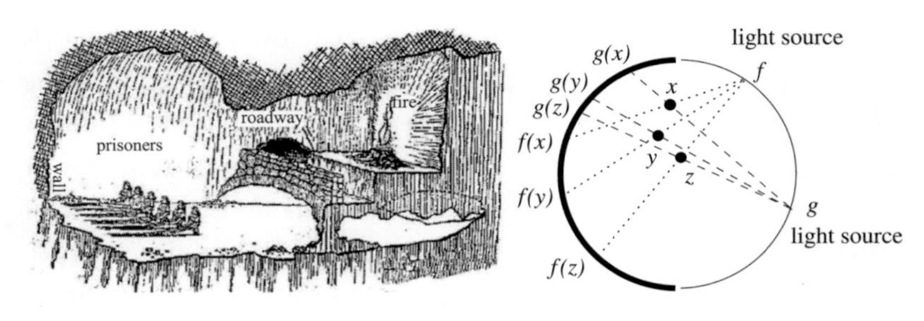

FIG. 2.13 Plato's cave allegory (left) and schematic (right).

as a deep principle in physics if not the whole content of physical reality. The hard-nosed reader can just take this second line with the meta-equation that I propose as an intriguing postulate independent of any philosophy, and note its mathematical consequences as explained later in this section.

What then is the fundamental nature of science? In the first approximation I consider the key part of this to be the concept of 'representation'. Aside from mathematics (logical thinking) there is, as mentioned above, the idea of theory and experiment, of something 'out there' and a 'measurement/image/representation' of it. Plato asked this question: how could prisoners confined all their lives to face the wall of a cave perceive the reality of the world behind them, such as a puppet show taking place on the roadway, from seeing only the shadows cast by a fire? See Figure 2.13. As a mathematician I have an answer to this, different from Plato.†

A representation of some mathematical structure X generally means a map from it to something considered concrete or self-evident such as numbers or matrices. What is important is not one but the set $\hat{X}$ of all such. Generally speaking this $\hat{X}$ has its own abstract structure and we can view each element $x \in X$ as a representation of this, that is we can take the view that $X = \hat{\hat{X}}$. Thus in mathematics one generally has a symmetrical notion in which

$$f(x) = x(f)$$

† Majid (1988a).

are two points of view, an element $f \in \hat{X}$ representing the structure of X and here measuring the value of a particular element $x \in X$, or the latter representing the structure of $\hat{X}$ and here measuring the value of a particular element $f \in \hat{X}$. In the context of Plato's cave then one should consider not only one angle of light, which creates one particular pattern of shadows as a representation of the real world X on the roadway, but the set $\hat{X}$ of all such representations corresponding to all possible angles $f, g, \cdots$, of the light source. In other words, allow the fire in Plato's cave to move around to give different sets of shadows. Here I consider one particular geometry of various point-like objects x, y, z on the roadway as the real structure X, the corresponding bits of the shadow in angle f are recorded as the locations $f(x), f(y), f(z)$ projected on the wall.† In some other representation g the shadows would be at $g(x), g(y), g(z)$, see Figure 2.13. One would be able entirely to reconstruct the reality X from knowledge of the images or 'measurements' in sufficiently many representations, certainly if you knew them for the whole set $\hat{X}$ of all representations. In fact two would typically be enough in the 2-dimensional schematic – just triangulate back to see where the dotted line that goes to $f(x)$ crosses the dashed line that goes to $g(x)$ and you know where in reality x is, similarly for y, z, except that you might also need to go smoothly from f to g and watch the shadows quite carefully at certain critical angles where they overlap each other, in order to keep track of which shadow goes with which object. Such 'tomography' seems all very reasonable but now consider: mathematics and the symmetrical nature of measurement as explained above means that *one could equally well say that the set of angles $\hat{X}$ of the light source was the 'real structure' and that each object x on the roadway was in fact a representation of $\hat{X}$*. A bit of the shadow $x(f)$ was the image of the angle f in this particular representation x. It would be quite hard to convince the prisoners that

† One could also consider X as the set of all possible locations of one object and x, y, z as trying out different locations in sequence, whereas we prefer to visualise the different elements of X all at once.

the real world that they longed to see behind them was nothing but a representation of the angles of the fire but from the mathematical point of view, if these are the only structures in consideration, it is equally valid. This reversal of point of view is an example of observer-observed duality. In a nutshell, while Plato's conclusion was that his cave was an allegory for a pure reality of which we see a mere shadow, our conclusion is exactly the opposite, that there is no fundamental difference between 'real' in this platonic sense and the world of shadows since one could equally well consider X as $\hat{\hat{X}}$, in other words as 'shadows' of what we previously thought of as shadows, and the latter as 'real' in the platonic sense.

The key question is, is such a reversed point of view physics or is this just a mathematical curiosity? The physical world would have to have the feature that the dual structures $\hat{X}$ would also have to be identifiable as something reasonable and part of it. This could be achieved for example by first of all convincing the prisoners to take $\hat{X}$ seriously, to think about its structure, to take on the view that x was a representation of this structure. Over time they might grudgingly allow that both X and $\hat{X}$ should be considered 'real' and that each represents the other. They would arrive in this way at a self-dual position as to what was 'real', namely the combined object $X \times \hat{X}$ where elements are a pair, one from each. This is often the simplest but not the only way to reach a self-dual picture. We propose this need for a self-dual overall picture as a fundamental postulate for physics:

> **First principle of self-duality:** A fundamental theory of physics is incomplete unless self-dual in a representation-theoretic sense. If a phenomenon is physically possible then so is its observer-observed reversed one.

We showed in Section 2.4 how this principle can be applied to quantum-mechanical algebras of observables and what it leads to there, namely to quantum groups and to constraints on the background geometry reminiscent of Einstein's equations.

One can also say this more dynamically: *as physics improves its structures tend to become self-dual in this sense*. This has in my view the same status as the second law of thermodynamics (that entropy always increases): it happens tautologically because of the way we think about things. In the case of thermodynamics it is the very concept of probability which builds in a time asymmetry (what we can predict given what we know now) and the way that we categorise states that causes entropy to increase (when we consider many outcomes 'the same' then that situation has a higher entropy by definition and is also more likely). In the case of the self-duality principle the reason is that in physics one again has the idea that something exists and one is representing it by experiments. But experimenters tend to think that the set $\hat{X}$ of experiments is the 'real' thing and that a theoretical concept is ultimately nothing but a representation of the experimental outcomes. The two points of view are forever in conflict until they agree that both exist and one represents the other.

It is not necessary in thermodynamics to understand the mechanisms really involved in proving the second law. In the same way we don't need to understand the reasons for the self-duality, but let us speculate on the mechanism anyway, for good measure (the reader can skip this paragraph quite easily). Note that we already explained in our discussion of Figure 2.1 that physics is in a sense dynamic, with theories merging into more general ones as physics evolves. I think a key element to the dynamic here is an 'urge' coming from the nature of being a physicist that structures should interact. One is not really happy with X and $\hat{X}$ as independent bits of reality. So long as they are both 'real' they should be part of some more unified structure. This creates a kind of 'engine' that could be viewed as driving the evolution of physics (in an ideal world rather than necessarily how it evolved in practice which will be more haphazard). Thus, along with the above principle is the dynamic urge to then find a new structure X_1 containing $X, \hat{X}$ as a basis for a more general theory. Experimenters will then construct representations $\hat{X}_1$ and we are back on the same path in a new cycle. Sometimes it can happen that a structure is 'self-dual', $\hat{X} \cong X$, in which case the explanation that one is really the other in a dual point of view means that we have no urgency to generalising the pair X, $\hat{X}$ to a new object. So for self-dual objects the 'engine' stalls and one could consider a theory of physics with such a structure particularly satisfying. For this to be possible the type of object that is $\hat{X}$ has to be the same type that is X, i.e. one has to be in a self-dual category (which is just a self-dual object at a higher up 'level' in a hierarchy of structures). This gives a mathematical formulation of the above principle, namely that:

> **Second principle of self-duality:** The search for a fundamental theory of physics *is* the search for self-dual structures in a representation-theoretic sense.

This self-duality is our proposed 'meta-equation' for the structure of physics. If you don't like the philosophy you can write here 'entails' in place of 'is'. Moreover, it is obviously an idealisation and hence is not going to be exactly what physics is, but it gives us something concrete to work with which could be used as insight. In physical terms it says that a complete theory of physics should have parts which represent each other hence allowing the reversal of roles of observer and observed, and in the self-dual case the reversed theory has the same form as the original one. The stronger interpretation of the principle here says that the reversed theory is indistinguishable from the original up to a relabelling of names of objects and associated units of measurement. The weaker one says that the reversed one may be some other theory but of the same categorical type. Another general remark is that there is a kind of 'fractal-like structure'. What I mean is that one can apply the self-duality at several levels, which in some sense are the same concept in different settings. Thus one can look for a self-dual category of objects, but within that one can look for and perhaps find an actual self-dual object, say some quantum group isomorphic to its dual. Within that object, assuming it has elements, one can look for and perhaps find an actual self-dual element invariant under a generalised 'Fourier transform', while going the other way the self-dual axioms of our first category could perhaps be viewed as self-dual objects at a higher level still where objects were axiom-systems.

To give a simple example, initially one might think in Newtonian mechanics that the structure of ordinary 3-dimensional space $X = \mathbb{R}^3$ was the 'real thing'. Its structure is that of an Abelian group (there is an addition law for vectors $\vec{\mathbf{x}} = (x, y, z)$) and this is used to define differential calculus, Newtonian mechanics and so forth. However, experimenters soon found that the things of particular interest were the plane waves, which mathematically are nothing other than representations of this additive law of X. Physicists think in terms of diagonalising a wave operator, but this operator is defined by differentiation and in fact the plane waves do more, they are functions

$$\psi_{\vec{\mathbf{p}}}(\vec{\mathbf{x}}) = e^{\imath\vec{\mathbf{x}}\cdot\vec{\mathbf{p}}}$$

where differentiation has a definite value labelled by a triple of real numbers $\vec{\mathbf{p}} = (p_x, p_y, p_z)$ called the momentum of the wave. The set of such waves itself forms an Abelian group $\hat{X}$ called 'momentum space'. It is again a 3-dimensional space and the addition law is that of momentum vectors. It was the nineteenth-century Irish mathematician and astronomer William Hamilton who realised that it was more natural to reformulate Newton's laws in a more symmetrical way in terms of both X and $\hat{X}$ (the combined space is called 'phase space' as we have seen in Section 2.4) and by now this is an accepted part of physics.† It is a 6-dimensional space for each particle in the system. In this point of view both X and $\hat{X}$ are equally real and represent each other. The 'reversal of view' expressed in the first principle is the normal notion of Fourier transform in which any reasonable function on position space can be expanded as a linear combination of plane waves with some coefficient for each $\vec{\mathbf{p}}$ – these coefficients then constitute a function on momentum space, which is the equally good Fourier transformed function. The self-duality of the combined

† He also invented the quaternion algebra mentioned in Connes' essay.

$X \times \hat{X}$ expresses that one is free to view the same plane wave above as $\psi_{\vec{\mathbf{x}}}(\vec{\mathbf{p}})$, reversing the roles of position and momentum.

The Fourier transform here is a change of basis from sharply peaked so-called 'delta functions' labelled by points in position space, to a plane-wave basis – both are used and provide complementary 'wave-particle' points of view on any phenomenon. According to the principle above, a theory of mechanics based just on point particles in position space would not be complete from a phenomenological point of view, one needs both. We can also look for self-dual objects. Indeed, certain special functions (called Gaussians) are self-dual under Fourier transform and play a fundamental role in physics. Finally, the dynamics, which is to say the time-evolution of the variables, is governed in a symmetric way by a 'Hamiltonian' function on the overall phase space. This is a second layer of the self-duality postulate, made possible when the paradigm is self-dual and in which dynamics is connected with self-duality of objects (and self-dual equations on them). Note that if we want X itself to be self-dual we need an inner product of some kind, which is to say the main ingredient which goes into such a Hamiltonian. And among Hamiltonians there is one of self-dual form which singles it out as a fundamental classical mechanical system in 'pole position', namely the one for a harmonic oscillator.

These ideas revolving around Fourier transform and the reversal of position and momentum could be called 'Born reciprocity' as mentioned towards the end of Section 2.2. They fit with our philosophy but our philosophy is much more general and allows us to extend them now to quantum gravity. Our addition to this story is that if one has both position and momentum $X \times \hat{X}$ as equally 'real' the next step is an urge to unify them into a single structure. Thus, Heisenberg proposed his algebra (2.10) for quantum mechanics in which $\vec{\mathbf{x}}$ and $\vec{\mathbf{p}}$ interact. But the resulting object X_1 in this case no longer respects our self-duality requirement – Heisenberg's algebra is not a group or quantum group of any kind, but we saw in Section 2.4 how this is possible in the form of the Planck-scale Hopf algebra by *also* introducing gravity in some sense. We see how the principle of self-duality is only a crude approximation for what is going on but leads one naturally to both quantum theory and gravity. We also saw at the end of Section 2.4 how quantum Born reciprocity appears in the context of quantum spacetime.

This explains how our previous Section 2.4 fits in with the general picture. Let us now step back and see what else it entails. What is the overall structure of the solutions of our 'meta-equation' which we proposed as what singles out the structure of physics from within mathematics (to first approximation)? If our simpler philosophical option above is right then *this is physics* (to first approximation). So to what extent does it resemble physics? Note that the 'equation' we have in mind is not about constructing one algebra or solving a differential equation but rather in first instance its indeterminates are axiom-systems for mathematical objects and the equation is that these axioms are self-dual in a representation-theoretic sense. Now the first thing to note about the meta-equation is that as an 'equation of physics' within the space of all mathematical axioms it has more than one solution, i.e. it predicts that theories of physics form more or less complete paradigms solving the self-duality constraint, that need be perturbed only when we wish to extend them to include more phenomena. It turns out that the meta-equation is also highly restrictive. There are not that many self-dual categories known. My own assessment of the mathematical scene is in Figure 2.14 with the self-dual ones along the central axis. According to our postulate above, the self-dual categories are 'sweet spots' as paradigms for complete theories of physics while straying off the axis breaks the self-duality and requires also to have the dual or mirror theory. This will not be exactly the structure of the historical paradigms of physics but is more like how the paradigms should have developed in an ideal world without the whims of history, according to our theory (which we only intend as a first approximation). The arrows in Figure 2.14 are functors or maps between different categories of structure whereby one type of structure can be generalised to another or conversely where the latter can be specialised to the former. These express how different paradigms

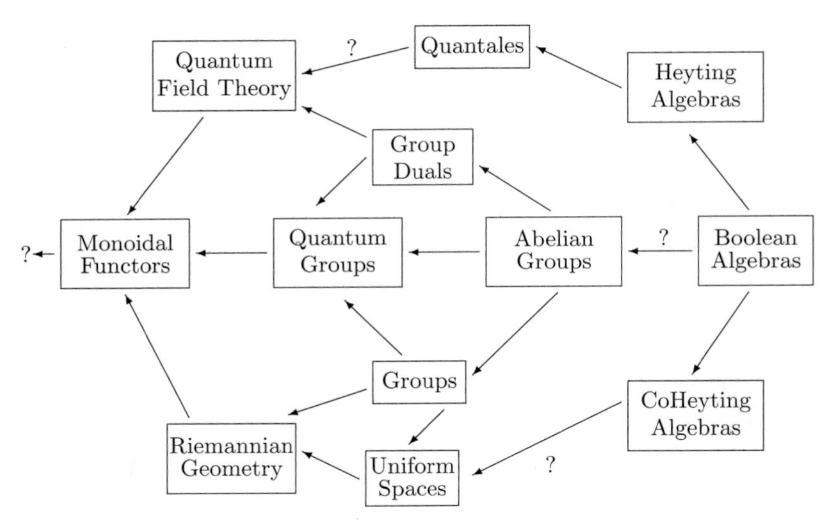

FIG. 2.14 Representation-theoretic approach to quantum gravity, from Majid (1988a). Arrows are functors between different paradigms expressed as categories of objects.

in physics in so far as we identify them can evolve one into the other or conversely remain as a special case. It is to be compared with our previous Figure 2.1.

On the axis we see the 'sweet spot' of Abelian groups (spaces with an addition law) as the setting for classical mechanics as discussed above. We discussed ordinary flat 3-dimensional space but the theory can also be applied to tori and to lattices (neither of these are self-dual) and to finite Abelian groups, which are self-dual. Abelian here means that the group law is commutative, which is why it is usually written as addition. Next, below the axis are nonAbelian groups, which as we explained at the end of Section 2.4 can be considered the first examples of curved spaces. Above the axis are their duals, which in this case are not groups but are defined using in fact the tools of quantum theory; they are the simplest quantum systems. So above the axis is quantum, below is gravity and they represent each other in some broad sense as illustrated in our discussion of Plato's cave. To unify them we need to put them into a single self-dual category namely that of quantum groups and there we can restore an observer–observed duality as we saw in Section 2.4. We see that quantum theory does not have observer–observed symmetry and nor does gravity, only in quantum gravity (by which we mean any form of unification) can this be restored and we propose that this being possible should in fact be a key feature of quantum gravity helping us to find it. The 'sweet spot' of quantum groups allowed certain toy models for this. But general curved spaces are not necessarily groups and real quantum systems are not necessarily group duals. For quantum gravity we therefore need a more general self-dual category and my proposal for this is shown in Figure 2.14 as the category of 'monoidal functors', as the next more general 'sweet spot'.

Before coming to that as the arguable 'end' of theoretical physics in the form of quantum gravity, let us look on the right side of Figure 2.14 and comment on its 'birth'. We take the view that the simplest theories of physics are based on classical Aristotelean logic or, roughly speaking, Boolean algebras. Boolean algebras have two operations $\cap, \cup$ standing for 'and' (conjunction) and 'or' (disjunction) of logical propositions, an operation $\bar{}$ for the negation of propositions, a zero element 0 for the always false proposition and a unit element 1 for the always true proposition, and some algebraic rules which you will already know if you think logically. For example, 'apples are round *or* it is not the case that applies are round' is always true,

$$A \cup \bar{A} = 1 \tag{2.12}$$

while 'apples are round *and* it is not the case that applies are round' is always false,

$$A \cap \bar{A} = 0. \tag{2.13}$$

One can also show such identities as

$$\overline{A \cap B} = \bar{A} \cup \bar{B}, \quad \overline{A \cup B} = \bar{A} \cap \bar{B} \tag{2.14}$$

which is known as 'de Morgan duality'. For example 'not the case that apples are round *and* square' = 'apples are not round *or* apples are not

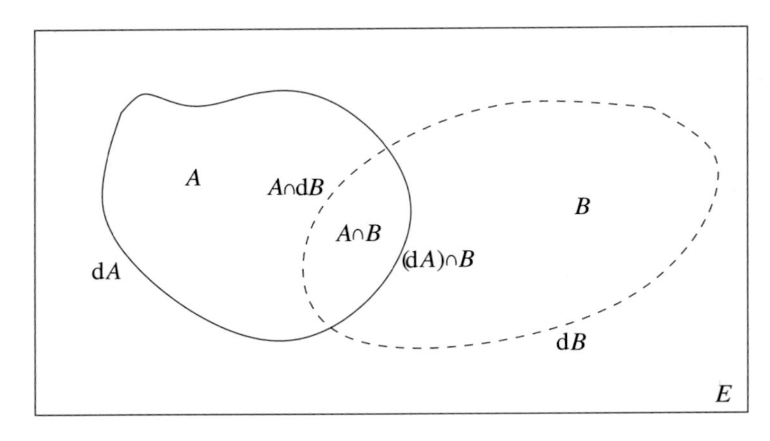

FIG. 2.15 Boundary of a set $\mathrm{d}A = A \cap \bar{A}$ behaves like differentiation for the product $\cap$. This is arguably the birth of geometry.

square'.† There is also a useful pictorial way of thinking about Boolean algebras. Suppose there is some universal set E of possible states of the system so that any propositions about it can be expressed as a subset or domain A inside E, namely that subset of states of the system where the corresponding proposition holds, see Figure 2.15. The conjunction of propositions corresponds to the intersection ('overlap') of subsets and the disjunction to union ('amalgamation') of subsets.

In topology one must say what it means to be 'in the neighbourhood' of a point, which means a specification of suitable families of 'open' and 'closed' subsets inside a space. All I want to say is that the collection of open and closed sets in any topological space forms a Boolean algebra so these stand exactly as the starting point for geometry at the level of topology. Also, we can think of a subset A of E as equivalent to a function on E with value 1 at every point that lies in

† It is also worth noting that in any Boolean algebra we have a partial ordering $A \subseteq B$ (standing for entailment or implication in propositional logic) defined as holding whenever $A \cap B = A$, which makes a Boolean algebra into a distributive complemented lattice (a point of view of relevance perhaps to posit approaches to quantum gravity).

the subset A and 0 elsewhere. Such functions form a 'classical algebra of observables' and making them noncommutative is 'quantum logic'. In this way, from a dual perspective, Boolean algebras are also at the starting point for quantum theory. This justifies the placement of Boolean algebras on the axis in Figure 2.14. One can also map any Boolean algebra over to an Abelian group defined by 'exclusive or'. In addition, there should be some sense in which de Morgan duality (2.14) can be considered as a representation-theoretic duality[†] but because I don't understand it fully, I have left a ? in the figure at this point.

Next, above the axis moving to Heyting algebras and beyond takes us into intuitionistic logic and ultimately into an axiomatic framework for quantum field theory. A Heyting algebra describes logic in which one drops the 'law of the excluded middle' (2.12). This generalisation is also the essential feature of the logical structure of quantum mechanics because in quantum theory a physical observable does not have to be either this or that value such as true or false, it can be in a mixed quantum state like Schrödinger's cat. *Less familiar* but dual to this is the notion of co-Heyting algebra and co-intuitionistic logic in which one drops the axiom (2.13) that the intersection of a proposition and its negation is always false. It has been argued by F. W. Lawvere[‡] and his school that this intersection $\mathrm{d}(A) = A \cap \bar{A}$ is like the 'boundary' of the proposition, and, hence, that these co-Heyting algebras are the 'birth' of geometry. This is depicted in detail in Figure 2.15. The boundary of the potato-shaped region A means literally the black line bounding A where the interior meets the exterior. Similarly $\mathrm{d}B$ is the dotted line shown. Now the part of the black line that lies inside B plus the part of the dotted line that lies inside A is clearly the boundary of the intersection $A \cap B$. Thus we see that

$$\mathrm{d}(A \cap B) = (A \cap \mathrm{d}B) \cup (\mathrm{d}A \cap B)$$

[†] Some inconclusive ideas for this are in Majid (2007).

[‡] Lawvere (1989).

an identity which can be proven properly in any co-Heyting algebra as proof that d expresses indeed 'the boundary' in the way that we have just visualised. The reader can compare with the rule for differentiating a product in high-school calculus. My long-term programme at the birth of physics is to develop this geometrical interpretation of co-intuitionistic logic further into the notion of metric spaces and ultimately into Riemannian or pseudo-Riemannian geometry.

Moreover, it has become apparent in recent years that there is a fundamental role of information in physics. If so, then this duality at the birth of physics should tie up with the other dualities above and play a fundamental role in quantum gravity itself. Certainly, it is something that is not possible in classical gravity: an apple curves space, but a not-apple which is perfectly equivalent as a concept in classical logic, does not curve space. Concrete matter curves space and this breaks de Morgan duality. But I envision that the symmetry could be restored in a theory of quantum gravity. When we say for example that one cannot have too many apples in a given volume because they form a black hole which would simply get bigger as you put more apples in (the right-hand slope in Figure 2.5), a person with an observer–observed reversed point of view on the same situation might say that space cannot be totally empty of not-apples (which is what that person might regard as the matter content). Indeed, the left slope in Figure 2.5 expresses quantum theory which says in its quantum field theory formulation that particles are constantly being created in particle–antiparticle pairs out of a vacuum and hence that space is never totally empty (this was the vacuum energy discussed at the end of Section 2.3). Note that a not-particle here as I envision it should not be confused with an antiparticle which (it is believed) curves space as much as a particle does. Yet the two are conceptually related in the 'Dirac sea' approach to matter particles such as electrons. This should all be resolved in quantum gravity. Thus, *I propose that observer–observed symmetry or representation-theoretic self-duality is tied up with a profound extension of de*

Morgan duality and hence with both time-reversal symmetry and charge conjugation.†

Finally, we come to the left-hand side of Figure 2.14 which, according to our principle of representation-theoretic self-duality, should be the setting for the 'end of physics' (as currently understood) in the form of quantum gravity. We need to go much beyond highly symmetrical group manifolds to general Riemannian (or pseudo-Riemannian) geometry on the one hand and beyond group duals to general quantum theory on the other. These are each categories of objects far from the self-dual axis in Figure 2.14 but again, I believe, in a somewhat dual relationship. Einstein's equation bridges the two and its consistent formulation should therefore be in a self-dual category general enough to contain both general geometry and general quantum theory and deformations of them that would be needed. The only thing is, which is perhaps one lesson from our analysis, it won't be exactly Einstein's equation in the form where it is supposed to equate something in classical geometry to the average value of a quantum object. Rather, the self-dual theory will likely involve a modification of both sides of the theory. We saw this already in Section 2.3 where for quantum spacetime Einstein's Special Relativity had to be a quantum group-based version – we should likewise expect a modification of both quantum theory and General Relativity in our new paradigm.

My proposal here‡ still stands on the table after two decades – what I think is the next most general self-dual category of objects after quantum groups in which to perform this unification, namely the category Mon_*. I suspect it would be impossible for me to describe it in any detail without too much pure mathematics, so suffice it to say that its main ingredients are (a) a fixed monoidal category – this means a category of objects for which there is some notion of

† Some of this was explored in Chapter 5 of Majid (1995) in the context of quantum random walks and 'coentropy'. Note that an antiparticle can be viewed as a particle going backwards in time.

‡ Eventually published in Majid (1991).

'product' of the objects, (b) another such category and (c) a map or 'functor' between the two. Let me stress that a category is nothing more than a clear way of speaking about what objects we consider and what 'maps' (morphisms) are allowed between them and how they are to be composed.† A functor between categories sends objects of one to objects of the other *and* morphisms of one to morphisms of the other. What I showed is that given such a triple, there was a notion of representation of it and that the set of such representations was itself a new triple in Mon_*.

These considerations of monoidal functors may seem far removed from quantum gravity but by now it is understood that they do in fact exactly solve quantum gravity in *three* dimensions and with point-source matter. It was already known by the end of the 1980s that most q-deformation quantum groups of the type that we associated in Section 2.3 with braid diagrams provide a way to measure how much a knot it knotted. Roughly speaking one draws the knot on a piece of paper with 'crossings' and reads it exactly as the circuit diagram of a simple 'quantum computer' as in Figure 2.7 with a braid operation or its inverse at each crossing. One can also view the system as a certain conformally invariant quantum field theory. Now, consider a 3-dimensional sphere with a knot drawn inside it. Thicken the knot to a tube cut out from the 3-dimensional sphere. Topologically this tube has the shape of a doughnut, so glue a doughnut onto our 3-dimensional sphere with knotted tube removed by mapping the surface of the doughnut to the surface of the knotted tube. The result is some 3-dimensional manifold‡ and every 3-dimensional manifold can be obtained in this way. In this way our measure of knottedness provided by the quantum group can be promoted to a measure of how complicated any 3-dimensional manifold is. When this is done for the right choice of quantum group, one obtains a number assigned to each 3-dimensional manifold which is nothing other than the weight

† A standard reference is Mac Lane (1974).

‡ The notion of a manifold is explained in more detail in the chapter by Penrose.

attributed in the 3-dimensional version of Einstein's formulation of gravity (the 'Einstein–Hilbert action') but *with* cosmological constant. If such ideas held in four dimensions we would use $(q-1) \approx 3 \times 10^{-61}$ according to (2.8) and our current values for the cosmological curvature scale.

Much of the mathematical part of the story above was developed in the late 1980s by mathematicians Jones, Reshetikhin, Turaev[†] and Viro. Vaughan Jones here received a Fields medal for his pioneering part. Much of the physical interpretation goes back to a translation in terms of 'Chern–Simons theory' by the physicist and Fields medallist Edward Witten.[‡] This part works for any q-deformation quantum group of a reasonable type but the part which is special to quantum gravity is a very particular choice of quantum group, namely a suitable 'q-Poincaré' one appropriate to 3 dimensions. It is not one of the standard Drinfeld-Jimbo q-deformations but obtained by a certain 'Drinfeld double' construction applied to one of them (in fact to $C_q[S^3]$ mentioned in Section 2.3). This Drinfeld double has a very natural interpretation in Mon_* as an example of the duality operation. Secondly, the guts of the construction above was a representation of a certain group $PSL_2(\mathbb{Z})$ of 'modular transformations' used to represent the gluing of the doughnut to the knotted tube. This is generated by something called a 'ribbon element' in the quantum group and, crucially, by the operation of Fourier transform expressed naturally in Mon_*. It does not correspond to a quantum group Fourier transform so it is outside of the quantum groups paradigm itself. Rather, the q-deformation quantum groups have braided versions and these are *self-dual* provided $q \neq 1$. Precisely because of this self-duality the (braided) Fourier transform becomes a matrix operator and this is the required generator of modular transformations.[§]

[†] Reshetikhin and Turaev (1990).
[‡] Witten (1989).
[§] Lyubashenko and Majid (1994).

We can also see the role in 3-dimensional quantum gravity of bicrossproducts and observable-state duality. This goes through the semidualisation construction mentioned towards the end of Section 2.4 – we consider a Poincaré quantum group $G\bowtie M$ where G is the momentum quantum symmetry group and M is the rotational quantum symmetry group. G^* is therefore the algebra of position variables. The position algebra here is a quantum analogue of the *local* 3-dimensional spacetime that exists at any point of a 3-dimensional manifold. In the dual model the roles of position and momentum are interchanged, so in the dual model $M\blacktriangleright\!\!\blacktriangleleft G^*$ is the Poincaré quantum group and G is the position algebra. The remarkable theorem is that in *either case, the position, momentum and rotations generate the same combined algebra*. This more fully expresses what we have alluded to as 'quantum Born reciprocity' at various points in the essay. Now, for most systems the semidualised system (I will just write 'dual' for brevity) where position and momentum are swapped, is a different system entirely. For example, the dual of (3-dimensional) classical gravity with cosmological constant $\Lambda \neq 0$ is a certain weak-gravity limit of quantum gravity with local structure *exactly* the 3-dimensional version of the bicrossproduct model flat quantum spacetime described in Section 2.3. Similarly, the semidual of quantum gravity with $\Lambda = 0$ is a certain classical theory with curved local spacetime. On the other hand, and here is the surprise, when you put in the specific q-deformation quantum groups you find

$$(\text{3-dimensional quantum gravity with } \Lambda \neq 0)^{\text{dual}} \sim \text{itself}$$

where I mean that it is self-dual up to an equivalence at a certain algebraic level.† The quantum groups under semidualisation in this last case have equivalent coproducts Δ (they are related by conjugation) and this induces an equivalence between the theories that they describe. Thus, our duality ideas applied to 3-dimensional quantum gravity not only provide insight into the key ingredients in its solution

† Details of this recent analysis will appear in Majid and Schroers (2008).

but its *self-duality* in the form of quantum Born reciprocity is a distinctive feature and requires both that spacetime be quantum *and* a nonzero value of the cosmological constant. In a sense *we have answered why there is a cosmological constant.*

Of course, 3-dimensional quantum gravity with point-particle matter is not the real thing. Even from the point of view of topology alone, the 4-dimensional case is the hardest of any dimension and one does not expect quantum groups themselves to be enough. According to Figure 2.14 we should look for more general objects in Mon_*. Recently, it was shown† that the construction of quantum field theory on curved spacetime *is* naturally formulated as the construction of a monoidal functor from the category of all pseudo-Riemannian geometries of a type suitable for quantum theory to the category of all quantum algebras of a suitable type. In other words, it cuts vertically across from the lower left to the upper left in Figure 2.14. Just as we did for quantum groups, one could now look for more general objects as a deformation of it (i.e. of both quantum theory of matter and classical gravity) within this paradigm Mon_* and perhaps under a self-duality type constraint. The deformation would induce the quantisation of gravity. Such an idea is pure speculation at the moment and life is probably far more complicated. But what all this shows is that the *principle of representation-theoretic self-duality* has non-trivial content capable of hinting at the mathematical structure of fundamental physics.

What comes as the next even more general self-dual category of objects? Undoubtedly something, but note that the last one considered is already speaking about categories of categories. One cannot get too much more general than this without running out of mathematics itself. In that sense physics is getting so abstract, our understanding of it so general, that we are nearing the 'end of physics' as we currently know it. This explains from our point of view why quantum gravity already forces us to the edge of metaphysics and forces us to

† Brunetti *et al.* (2003).

face the really 'big' questions. Of course, if we ever found quantum gravity I would not expect physics to actually end, rather that questions currently in metaphysics would become part of a new physics. However, from a structural point of view by the time we are dealing with categories of categories it is clear that the considerations will be very general. I would expect them to include a physical approach to set theory including language and provability issues such as those famously raised by the logician Kurt Gödel in the 1930s.

2.6 RELATIVE REALISM

I have explained in the last section a certain mathematical model of physics and we have seen a certain fit with the different paradigms of physics not necessarily historically (I am *not* making a sociological case) but in terms of its logical structure. If anything like this self-duality constraint *is* physics, then as we search deeper and deeper towards an ultimate theory of physics we are in fact rediscovering our scientific selves, as the self-duality has its roots in the very nature of science as an interaction between theory and experiment. If physics was always supposed to inform the rest of society about the 'nature of reality', what would such a thesis be informing us?

I think the resulting vision of reality that I shall try to enunciate now is one which need not throw out the 'hardness' of science, but the nature of this reality is nevertheless closer to the reality experienced by pure mathematicians than the mind-set usually adopted by working scientists in the West. I formulated this about 20 years ago under the term 'relative realism'.† I should say that I was not born a pure mathematician, it was always my vocation to study theoretical physics, quantum theory etc. and I always disliked pure mathematics as too unmotivated. But the problem of quantum gravity for the reasons above made me realise in the middle of my PhD that a pure mathematician's mind set was the correct one. Moreover, I do not wish to imply that pure mathematicians typically agree with me on

† Majid (1988a).

relative realism; most in fact take a more platonic view on mathematical reality, if they choose to think about such things at all. But it's my experience doing pure mathematics that I draw upon.

What I think a pure mathematician does understand deeply is the sense in which definitions 'create' mathematical reality. You define the notion 'ring' and this opens up the whole field of Ring Theory as a consequence of the axioms of a ring (a ring is a set with addition and multiplication laws, compatible with each other). Somebody invented the axioms of Ring Theory and if you don't like them you are free to define and study something else. In this sense mathematicians don't usually make a claim on 'reality'.

And yet, in some sense, the axioms of rings were there, waiting to be discovered by the first mathematician to stumble upon them. So in that sense mathematical reality is there, waiting to be discovered, and independent of humans after all.

Are these two statements contradictory? No, provided one understands that all of mathematics is similarly layered as a series of assumptions. Ultimately, if all of mathematics could be laid out before you (there are some technical issues there but I consider them inessential) it would be by definition all things that could be said within a set of assumptions known as being a mathematician (presumably some form of logic and certain rules as to what it entails to make a proof in any reasonable sense). So even the totality of the mathematical reality being uncovered by mathematicians is based on certain implicit axioms. In between this extreme and the example of Ring Theory are a whole hierarchy of assumptions that you could accept or not accept. Not accepting the associative law of multiplication takes you to a more general theory than Ring Theory (namely to the theory of 'nonassociative rings' by which is meant not necessarily associative rings). At that level there are facts, constraints, the hard reality of nonassociative Ring Theory, and it includes as a sub-reality the associative special case. And so on, dropping each of the assumptions of Ring Theory takes us to more and more general structures until we are studying very general things, such as the theory of Sets (at least a ring should be a set). And if one does not want to assume even a set, one is doing very general mathematics indeed.

Here we see a point of view in which it's very much up to you what you want to assume and study, but whatever you assume puts you on a certain 'level' of generality and this is the level of mathematical reality that you are experiencing as you proceed to research. Arguably the more general level is the more 'real' as the least is put in by an assumption, but equally convincing, the less general level is the more real as having 'real substance'. What we see here is *two points of view* on reality, on what really matters. If you are 'in the room' of Ring Theory looking around you see the nuggets of reality and hidden treasures that are Ring Theory. You might look out through the doorway and wonder what lies outside but you don't perceive it directly. On the other hand if you step outside you see the room of Ring Theory as one of many rooms that you might enter. You don't directly perceive the hidden treasures of Ring Theory but you are aware that you could if you went inside the room. Rather, the room itself is one of the nuggets of reality in your world at this 'level'. This is part of the paradox between the general and the specific that inspires pure mathematics research. We already mentioned at the start of Section 2.5 that it's often much easier to prove a general theorem that applies to a whole class of objects than to prove the same thing for a specific object. Conversely, it is the specific examples that ultimately make a general theorem of interest. A pure mathematician does not take one view or another but sees all the structure of (overlapping) nested boxes in its entirety as of interest, and recognises the choice in where to think within it.

In short, what is usually called reality is in fact a confusion of two mental perceptions. Wherever we are in the world of mathematics we have made a certain number of assumptions to reach that point. If we put them out of our mind we perceive and work within the 'reality' that they create. If we become self-aware of an assumption and consider it unnecessary we transcend to a slightly more general reality created by the remaining assumptions and look back on the aspects of reality created by the assumption that we became aware of as an arbitrary choice. Thus what you consider real depends on your attitude to a series of choices. This is a model of reality that is adapted to the fundamental nature of knowledge, achieves the *minimal goal* of explaining how we perceive it and also has at its core notions of consciousness or self-awareness and free will. These last are arguably missing ingredients of any quantum-gravity theory as we discussed in Section 2.4. We don't need to speculate further as to what these 'really' are in order to have an operationally useful philosophy but we do put them in pole position. In particular, the reality of 'Schrödinger's cat' in Section 2.4 is relative to your point of view, depending on whether you are in or out of the room with the cat.

Next, everything I said above could also be applied in everyday life. The example I like to give is playing chess. If you accept the rules of chess then you are in the reality of chess, you can experience the frustration and anguish of being in a check-mate. But if you step back and realise that it's just a game, the rules of which you don't have to take on, your anguish dissolves and you transcend to a more general reality in which the fact that chess is possible is still a reality but not one that you are immersed in. It is a nugget of reality at a higher level, the reality of board games for example, alongside other board games of interest for their rules rather than for actually playing. This reminds me a little *but only a little* of Buddhism. I am not an expert on Buddhism but its central tenet surely is that suffering, anguish, can be transcended by thought alone (meditation). In Tibetan Buddhism, by stages of contemplation, one progresses to greater and greater levels of enlightenment in which successive veils are lifted, which I understand as proceeding up to higher levels of generality in which more and more assumptions are become aware of as optional. In all forms of Buddhism this appears as the central role of compassion as a guide to dealing with others. In mathematics it appears as a compassion towards other researchers who have taken on different axioms. Ring theorists don't fight with group theorists about who is 'right'. Neither

FIG. 2.16 Is this a rocking chair? What does it mean to say that an actual chair stands before you? 'Correalistic Rocker', Friedrich Kiesler. Courtesy of the manufacturer Wittmann, Austria.

is 'true' and likewise Buddhism generally rejects our usual concrete reality or *samsara* as an illusion.†

Now returning to our ideas, let us illustrate them a little further at this level of everyday life; let us ask what does it mean to say that here in front of me is a chair? This is something solid and concrete, surely there is nothing relative about this. From an operational point of view it's enough to be able to say how someone could tell if there was a chair here. We could try to specify a recognition algorithm, but this is in fact a quite difficult task as anyone involved with AI will testify. Suffice it to illustrate some of the difficulties in Figure 2.16.

† But there are also fundamental differences – relative realism is very far from Buddhism. In Buddhism what is truly 'real' is what is left after letting go of all assumptions, even the very notions of 'is' and 'is not'. This is why, in the Zen tradition, the truth can only be seen in the smile that occurs when perceiving the contradiction in a Zen riddle. In this way Buddhism claims to avoid nihilism. In mathematical reality there is likewise something that remains after letting go of all assumptions, but it's just the entirety of the fine structure of all of mathematics. In Buddhism the attitude to *samsara* is pejorative, as an illusion it is to be dismissed. In relative realism the boxes that you can find yourself in *are* all there is, they are the nuggets of reality, the heart and soul of science, the whole point so to speak and very much to be celebrated. The key point is that *just because they are the product of assumptions does not make them arbitrary and worthless.* That would be like saying that pure mathematics was arbitrary and worthless which is not really the case.

Is the item shown a (rocking) chair as claimed by its designer and its Austrian manufacturer of surrealist furniture? It does not have legs, for example. At the end of the day I think the best definition which we in fact all use is: *a chair is whatever you and I agree to be a chair when you and I use this word in a particular communication*. To the extent that we agree, that is the extent to which there is a chair in the room in the case of the figure. It is not even a matter of function over form, it's a matter of the 'handshaking' agreement that is involved in any act of communication and the choice to define something like this and then to use that definition.

In relative realism then we sort of 'swim' in a sea of definitions of which a crude approximation might be the choices to use a few hundred thousand words or concepts. Some who use more have a richer experience of reality while I believe the minimum for getting by in everyday life is judged by linguists at about 600 which would be a sterile and angst-ridden experience. But does something exist if we don't have a word for it but someone else does? It exists for them and not for us, provided one understands that we are using the word 'exist' here in a certain way. This is why it is relative realism; that there should be a single answer is the old materialism that it replaces. Note that in practice we would also need to take account of the baggage of a whole sea of other assumptions that would be likely in practice to play the role of a missing word. Moreover, while one could say that what replaces mathematical reality in everyday life is crudely approximated by linguistic reality, I don't really mean just the words but nonverbal concepts too, as well as the entirety of their inter-relations to greater and lesser extents. This is all supposed to be smoothed over and integrated to give an overall 'perception' of solid reality. In mathematical terms, we have talked above about letting go of assumptions, transcending to more and more general points of view. But the other limit of this 'tree of knowledge' is the limit of finer and finer assumptions that one might be more and more dimly aware of taking on, a limit which I once called the 'poetic soup'. It's the opposite of abstractification, so let us call it a model of direct perception. Moreover, and this is key, our feeling of solidity or grounding exists to the extent that others use the same definitions or (in the limit) perceive the same. It is the extent to which we are 'in touch' with the real world. I am reminded of spy movies from the 1960s in which the first thing before starting off on a mission is for our heroes to synchronise their watches. I think that as we communicate we synchronise our terms and concepts to the extent that we communicate successfully and to that extent we then agree on what we are just doing.

I suppose sometimes the communications will be literal ones as events in spacetime and there are issues similar to those that already exist in quantum mechanics in the context of the EPR thought-experiment. On the other hand, by 'communication' in relative realism here we don't necessarily mean a sharp event, we intend any kind of influence or interaction to a greater or lesser extent by which concepts are transferred and synchronised. Moreover, communication does not necessarily entail a conscious entity doing it. A synonym for each side of a communication could be 'awareness' or 'to be conscious of' and maybe this is all there actually is to consciousness from an operational point of view. This would be analogous to the way that the measurement postulate in quantum theory shelves the issue while remaining operational. Note that 'information' defined as that which is communicated is entering here into the discussion and in physics this links back to thermodynamics and, these days, to gravity.

And could such a scheme ever reach the clarity and precision of the physical sciences, where we agree on many things to 13 decimal places so to speak? Part of the answer to this is that the assumptions going into it are far more primitive and uniformly accepted than, say, the rules of chess. They are by definition more related to the physical world and might include, for example, assumptions formed as we first learned to walk. The second part of the answer is the scientific method of which we made a crude approximation in Section 2.5 in the form of 'self-dualism'. Something like that could single out sharp structures and substructures to be the 'hard sciences' while the realities of chess and of chairs are more flexible.

BIBLIOGRAPHY

Beggs, E. J. and Majid, S. (2001) Poisson-Lie T-duality for quasitriangular Lie bialgebras, *Commun. Math. Phys.* **220**, 455–88.

Brunetti, R., Fredenhagen, K. and Verch, R. (2003) The generally covariant locality principle – A new paradigm for local quantum physics, *Commun. Math. Phys.* **237**, 31.

Connes, A. (1994) *Noncommutative Geometry*, Academic Press.

Drinfeld, V. G. (1987) Quantum groups, in *Proceedings of the ICM*, American Mathematical Society.

Einstein, A. (1950) *The Meaning of Relativity*, Princeton University Press.

Gomez, X. and Majid, S. (2002) Noncommutative cohomology and electromagnetism on $C_q[SL_2]$ at roots of unity, *Lett. Math. Phys.* **60**, 221–37.

Heisenberg, W. (1927) "Über den anschaulichen Inhalt der quantentheoretischen Kinematik und Mechanik", *Zeitschrift für Physik* **43**, 172–198. English translation in J. A. Wheeler and H. Zurek, *Quantum Theory and Measurement*, Princeton University Press, 1983, pp. 62–84.

Lawvere, F. W. (1989) Intrinsic boundary in certain mathematical toposes exemplify 'logical' operators not passively preserved by substitution. Preprint, University of Buffalo.

Lyubashenko, V. and Majid, S. (1994) Braided groups and quantum Fourier transform, *J. Algebra* **166**, 506–28.

Mac Lane, S. (1974) *Categories for the Working Mathematician*, Springer.

Majid, S. (1988a) The principle of representation-theoretic self-duality; later published *Phys. Essays* **4**, 395–405.

Majid, S. (1988b) *Non-commutative-geometric Groups by a Bicrossproduct Construction* (PhD thesis, Harvard mathematical physics); Hopf algebras for physics at the Planck scale, *J. Classical and Quantum Gravity* **5**, 1587–606; Physics for algebraists: non-commutative and non-cocommutative Hopf algebras by a bicrossproduct construction, *J. Algebra* **130**, 17–64.

Majid, S. (1991) Representations, duals and quantum doubles of monoidal categories, *Suppl. Rend. Circ. Mat. Palermo, Series II* **26**, 197–206.

Majid, S. and Ruegg, H. (1994) Bicrossproduct structure of the κ-Poincaré group and noncommutative geometry, *Phys. Lett. B* **334**, 348–54.

Majid, S. (1995) *Foundations of Quantum Group Theory*, Cambridge University Press.

Majid, S. (2005) Noncommutative model with spontaneous time generation and Planckian bound, *J. Math. Phys.* **46**, 103520.

Majid, S. (2006a) Algebraic approach to quantum gravity III: noncommutative Riemannian geometry, in *Mathematical and Physical Aspects of Quantum Gravity*, ed. B. Fauser, J. Tolksdorf and E. Zeidler, Birkhauser.

Majid, S. (2006b) Algebraic approach to quantum gravity II: noncommutative spacetime, to appear in *Approaches to Quantum Gravity*, ed. D. Oriti, Cambridge University Press.

Majid, S. (2007) Algebraic approach to quantum gravity I: relative realism. Preprint.

Majid, S. and Schroers, B. J. (2008) q-Deformation and semidualisation in $2+1$ quantum gravity, I. *In preparation.*

Reshetikhin, N. Yu. and Turaev, V. G. (1990) Ribbon graphs and their invariants derived from quantum groups, *Commun. Math. Phys.* **127**, 1–26.

Witten, E. (1989) Quantum field theory and the Jones polynomial, *Commun. Math. Phys.* **121**, 351.

3 Causality, quantum theory and cosmology

Roger Penrose

3.1 SPACETIME STRUCTURE

Einstein's general theory of relativity gives a mathematical description of space, time and gravitation which is extraordinarily concise, subtle and accurate. It has, however, the appearance of being concise only to those who are already familiar with the mathematical formalism of Riemannian geometry. To someone who is *not* familiar with that body of mathematical theory – a theory which, though remarkably elegant, is undoubtedly sophisticated, and usually becomes extremely complicated in detailed application – Einstein's General Relativity can seem inaccessible and bewildering in its elaborate structure. But the complication and sophistication lie only in the details of the formalism. Once that mathematics has been mastered, the precise formulation of Einstein's physical theory is, indeed, extremely compact and natural. Although a little of this formalism will be needed here, it will be given in a compact form only that should be reasonably accessible.

The mathematical theory of Riemannian geometry applies to smooth spaces of any (positive whole) number N of dimensions. Such a space $\mathcal{M}$ is referred to as an *N-manifold*, and to be a *Riemannian* manifold it must be assigned a *metric*, frequently denoted by $\mathbf{g}$, which assigns a notion of 'length' to any smooth curve in $\mathcal{M}$ connecting any two points a, b. (See Figure 3.1.) For a strictly Riemannian manifold, this length function is what is called *positive definite* which means that the length of any such curve is a positive number, except in the

On Space and Time, ed. Shahn Majid. Published by Cambridge University Press.

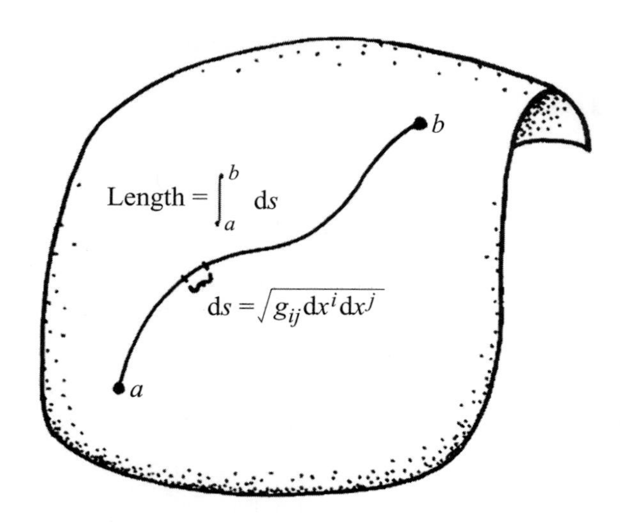

FIG. 3.1 The metric tensor **g** provides a notion of *length* along a curve connecting two points a and b.

degenerate situation when $a = b$ and the curve shrinks to a point, for which the length would be zero. Einstein's theory, however, uses what is more correctly called *pseudo*-Riemannian geometry[†] to describe spacetime, and although the mathematical formalism is almost identical with that for the Riemannian case, there is the important difference that this condition of positive definiteness now fails. A curve connecting distinct points a and b in pseudo-Riemannian geometry can sometimes become zero for *non*-degenerate curves – which is physically reasonable because the interpretation of this pseudo-Riemannian length is the *time* experienced by some physical particle between events a and b (the particle being thought of as providing a kind of 'clock') whose *world-line* stretches from a to b. (The world-line describes the particle's history between the events a and b.) When the particle is a free photon (or other massless entity), the time that it experiences is *zero* even when a and b are distinct. This is an extreme example of the *time-dilation* effect of relativity theory, in the limit

[†] The term "semi-Riemannian" has also sometimes been used to describe this situation, but this strikes me as a less appropriate terminology.

of when a 'clock' (here a photon, or other massless particle) actually travels at the speed of light. I shall return to this important issue later.

In physical terms, it is fortunate that Nature provides us with an extremely accurate and unambiguous notion of a 'clock'. This is based on two of the most fundamental principles of twentieth-century physics, as embodied in two famous formulae for the energy E of a system, due respectively to Max Planck and Albert Einstein:

$$E = h\nu \quad \text{and} \quad E = mc^2.$$

Accordingly (eliminating E between these two equations), any particle of mass m is, in a sense, a very precise 'clock' which ticks away with a frequency ν that is proportional to this mass

$$\nu = m\frac{c^2}{h},$$

where h and c are Planck's constant and the speed of light, respectively. In practice, it would not be possible to make a usable clock out of just a single particle, and more elaborate structures involving numerous interacting particles need to be used. But the ultimate principle is the same. It is the robustness of the *mass* concept, and of the interactions that are involved in determining that mass, which are responsible for the spacetime pseudo-metric being such a precisely physically determined measure.

Physical *distance* measurements are best thought of as being determined by time measurements. The speed of light c translates time units into distance units. It is customary for physicists to think of time and distance measurements to be basically interchangeable, via the speed of light (years to light-years, seconds to light-seconds, etc.). However, time measurements are now so much more precisely determined than distance measurements that the metre is now *defined* so that there are precisely 299 792 458 of them to a second (so a light-second is now exactly 299 792 458 metres)!

To emphasise the precision that time measurements have in observational physics and also in modern technology, we may call to mind GPS devices, which can determine a location on the ground to less than a metre, via the precise *timing* of signals from Earth satellites (which determination, incidentally, must call upon Einstein's General Relativity to obtain such locational precision). Even more impressive is the analysis† of the double-neutron-star system PSR 1913+16, one of whose components is a pulsar emitting electromagnetic pulses, dynamically precisely determined from the rotation period of the body of the star. These have been tracked on Earth for about 30 years, giving an agreement between observation and theory (Einstein's General Relativity, that is, with implications ranging from Kepler's elliptical motions, via Newton's corrections from universal gravitation, and from more detailed corrections to Newton's theory from Einstein's General Relativity, with final corrections involving energy-loss effects due to the emission of gravitational waves) giving an overall precision of one part in about one hundred million million (10^{14}) over 30 years, i.e. to an error of within only about one hundred-thousandth of a second, over that entire period! This is an extraordinary confirmation, not only of the accuracy of Einstein's theory, but also of the agreement between the dynamical behaviour of large-scale bodies and of the timing provided by terrestrial (atomic and nuclear) clocks, these depending ultimately on the quantum-mechanical principles referred to above.

We need to look a little more carefully at the mathematical description of the passage of time, according to relativity theory. Time, being described in a way similar to the determination of distance in Euclidean geometry, arises from a version of the Pythagorean theorem, at an infinitesimal scale, so it is expressed as the square root of some quadratic expression (the length of a hypotenuse being the square root of a sum of squares of the individual component side-lengths, in the Pythagorean theorem). In Riemannian geometry, we use general coordinates $(x^1, x^2, x^3, \ldots, x^N)$ to label the different points in the manifold

† This earned Joseph Taylor and Russell Hulse the 1993 Nobel Prize for physics. For a description of this work, see Taylor (1998).

$\mathcal{M}$ (in a local[†] region – where the upper indices simply distinguish the different coordinates and do *not* indicate powers of a single quantity 'x'). Then we have the infinitesimal length formula

$$ds^2 = g_{ij} dx^i dx^j,$$

where Einstein's *summation convention*, according to which a summation over repeated indices is assumed, is being adopted. The infinitesimal length interval is ds, so the *finite* length of a curve between points a and b is given by the formula (Figure 3.1)

$$\int_a^b ds$$

where the integral is to be taken along the curve. The N^2 quantities g_{ij} (each of i and j running over $1, 2, \ldots, N$) are defined at each point in $\mathcal{M}$ and provide the components of the *tensor* **g** of *valence* $\begin{bmatrix} 0 \\ 2 \end{bmatrix}$, where the '2' refers to the number of lower indices, there being no upper ones in this case. In fact, there are only $N(N+1)/2$ *independent* components, owing to the symmetry $g_{ji} = g_{ij}$. In Riemannian geometry, the matrix of components (g_{ij}) is positive definite, whereas in Einstein's General Relativity, it is to have the Lorentzian signature $(+---)$ in the sense that the time direction contributes positively to the Lorentzian distance measure while the space directions contribute negatively. We are concerned with $N = 4$, so that in this case there are 10 independent components.

This choice of signature has the implication that there is a *light cone* of directions[‡] at any point p of the spacetime along which

[†] The word 'local', as used here, refers to some finite or infinite (but not infinitesimal) region which need not extend to the entire manifold. (Such a region would be an open set, in technical terms; see for example Penrose (2004, §7.4) for a brief introduction to the notion of 'open set'.)

[‡] My own favoured terminology would normally have been to refer to this 'light cone of directions' at p as the '*null* cone' at p, and to reserve 'light cone' for the family (or locus) of spacetime light rays through p. Technically, this null cone is a structure within what mathematicians refer to as the *tangent space* at p, and is therefore a structure in the tangent space rather than in the spacetime itself. See Penrose (2004, §§17.6–9).

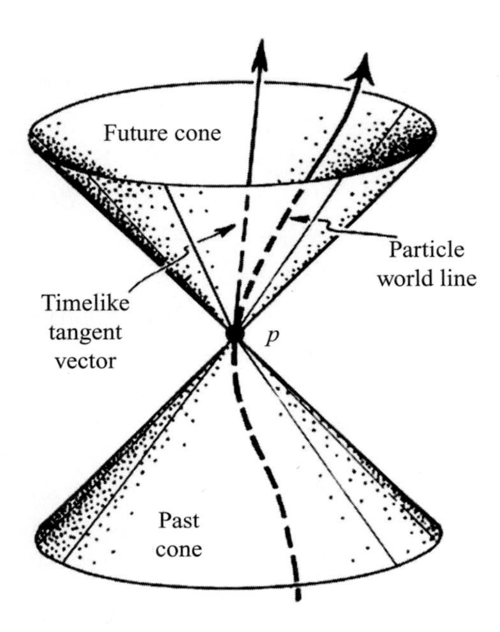

FIG. 3.2 The light cone at a spacetime point p. The world-line of a massive particle has *timelike* tangent directions, which necessarily lie within these cones. The tangent to a massless particle would point along the cone.

the length measure is *zero*. See Figure 3.2. The complete light cone consists of two components, connected to each other only at the point p, which is the common vertex of each. One component is the *future* light cone, which represents the history of a flash of light which originates at the event p, and then spreads out throughout a region of $\mathcal{M}$, to the future of p. The other is the *past* light cone, which represents the history of a light flash which converges on the event p from a region to the past of p in $\mathcal{M}$. For directions *within* the light cone, the squared interval ds^2 is positive, so we can take the positive square root to give a positive measure ds for the (infinitesimal) interval of time measure into the future away from p. Such a direction is referred to as *timelike* and the world-lines of massive particles are always timelike (in the sense that their spacetime directions are everywhere timelike). However, for directions at p that point to regions *outside* the light cone, ds^2 is *negative*, so that the (infinitesimal) interval ds would have to be imaginary. Such directions are referred to as *spacelike*.

This makes sense when we consider the demands of relativity, where we interpret $\int ds$ as a measure of *time*. The world-line of an ordinary massive particle which contains the event p must indeed lie within this cone, reaching p from within the past light cone and emerging from p within the future light cone. The world-line of a free photon (or other massless particle) would be taken to lie *on* the light cone, with portions on the past light cone and future light cone joined at p. But spacelike directions are not permitted for actual physical particles, since these would correspond to motions faster than the (local) speed of light. These features express the standard causality requirements of relativity theory, according to which signals and the motions of physical particles are constrained by the light cones: communication is restricted so that information cannot propagate by superluminary means (at least in the local sense), so the imaginary value for the 'time interval' ds is now consistent with this because it is never measured by actual classical particles. This *causal* role of the light cones is, in some sense, even more fundamental than their relation to ideal clocks in relativity theory. We shall be seeing the significance of this later.

In order to formulate Einstein's theory, we need some further tensor quantities, all of which are mathematically determined by the metric tensor $\mathbf{g}$. First, we construct the *inverse* $\mathbf{g}^{-1}$ of $\mathbf{g}$, which has valence $\left[\begin{smallmatrix}2\\0\end{smallmatrix}\right]$, where the matrix of components (g^{ij}) of $\mathbf{g}^{-1}$ is the inverse of the matrix of components (g_{ij}) of $\mathbf{g}$ at each point in $\mathcal{M}$. Remember that the tensor $\mathbf{g}$, like all tensors in tensor analysis, can vary from point to point, so that we can differentiate it in the sense of differential calculus to obtain new objects. In particular, from $\mathbf{g}^{-1}$ and derivatives of $\mathbf{g}$, we can construct a somewhat more complicated, but very fundamental, tensor quantity referred to as the Riemann–Christoffel *curvature* tensor $\mathbf{R}$ which has valence $\left[\begin{smallmatrix}0\\4\end{smallmatrix}\right]$. One of the fundamental roles of $\mathbf{R}$ is to tell us when $\mathcal{M}$ is the *flat* Minkowski space $\mathbb{M}$ of special relativity. The condition $\mathbf{R} = 0$ is, indeed, necessary and sufficient for $\mathcal{M}$ to be *locally* the same as $\mathbb{M}$.

In Einstein's theory, the gravitational field is not described as a 'force' in the ordinary way, similarly to the descriptions in terms of forces which apply to all other physical interactions, because in a freely falling frame of reference the 'gravitational force' is not felt at all. (This is somewhat ironic, considering that it was Newton's description of gravitation in such 'force' terms that set modern science on its course, establishing the Newtonian framework for physics which still holds today!) This *principle of equivalence* between the effects of gravitation and of acceleration allows the gravitational force at any one point to be eliminated by passing to a freely falling frame of reference, but the spatial variation of this force – i.e. the *tidal effect* – cannot be so eliminated, and it is this which is described in terms of the spacetime curvature **R**.

The term 'valence' in tensor analysis, is chosen from its analogy with chemical bonding, as tensors can be joined together in a form of multiplication, whereby upper tensor indices and lower ones can be connected with one other by a process of 'contraction' (summation over repeated indices, one upper and one lower), or not, as desired. An example of this, we can join the $\left[\begin{smallmatrix}2\\0\end{smallmatrix}\right]$-valent inverse metric $\mathbf{g}^{-1}$ and the $\left[\begin{smallmatrix}0\\4\end{smallmatrix}\right]$-valent **R** with two contractions, to form the $\left[\begin{smallmatrix}0\\2\end{smallmatrix}\right]$-valent *Ricci* tensor **P**. Now the *source* of gravitation is *mass* which, according to Einstein's famous $E = mc^2$ above, is in effect *energy*. For a continuous source distribution, we must think of energy *density* (the energy per unit volume) which, in relativity theory, is one typical component of a $\left[\begin{smallmatrix}0\\2\end{smallmatrix}\right]$-valent quantity **T**, called the *energy-momentum* (or *stress-energy*) tensor. In his first attempts at a general-relativistic gravitational theory, Einstein tried equating **T** with **P** (with a constant multiplier), but he later realised that this would not work, because **T** has to satisfy a differential relation referred to as a 'conservation law' which I shall write as $\nabla \cdot \mathbf{T} = 0$ (this quantity being a certain contracted derivative of **T**). The trouble with Einstein's original suggested identification of **T** with **P** is that generally $\nabla \cdot \mathbf{P}$ does *not* similarly vanish. But subsequently he realised that another $\left[\begin{smallmatrix}0\\2\end{smallmatrix}\right]$-valent tensor **G**, that can be directly constructed from $\mathbf{P}$ by a simple operation of 'reversing its trace', rather remarkably *does* satisfy $\nabla \cdot \mathbf{G} = 0$. Einstein's gravitational field equations then took the very elegant form (where G is Newton's gravitational constant, the minus sign merely coming from the particular sign conventions that I and many others happen to adopt):

$$\mathbf{G} = -8G\pi\mathbf{T},$$

and it is *this* simple equation which provides us with the astounding accuracy that I referred to above.

There is, however, a remarkable coda to this extraordinary story. Einstein had observed that there is one (and only one) slight modification that he could make to the above equation which leaves all the principles underlying the theory unaffected, except that in the limit of weak fields it does not quite go over to Newton's theory, but instead there is an overall cosmic repulsion that would not make itself noticed until cosmological scales are encountered, namely the introduction of a tiny constant number Λ (called the *cosmological constant*) that could be introduced, multiplying the metric $\mathbf{g}$, as an additional term, so that his equation now becomes

$$\mathbf{G} + \Lambda\mathbf{g} = -8G\pi\mathbf{T}.$$

Einstein originally proposed this modified form of his equation in 1917, one year after his original publication of his field equation, in order to be able to have a solution in which the Universe is *static* (i.e. unchanging with time) in its overall structure, which he had his reasons for believing should be the case. But when, in 1929, Edwin Hubble provided convincing evidence that the Universe is in fact expanding, Einstein changed his mind, and it is said that he regarded the introduction of Λ as his 'greatest mistake'.

However, once introduced, the possibility of a nonzero Λ did not 'go away', and it became a matter of serious study in all major texts on cosmology. Cosmological observations remained consistent with $\Lambda = 0$ until about 1998, when measurements of distant supernova explosions[†] began to indicate that there is indeed a rate of cosmic expansion that is consistent with, and explained by, the presence of this cosmological term with Λ having a tiny positive value! Subsequent observations have provided additional support for this. Many cosmologists seem to be favourably disposed to the possibility that such a 'Λ' might actually be varying, and in accordance with such a possibility, have re-christened the cosmological term as 'dark energy'. But the notion of a 'varying Λ' presents various theoretical difficulties of its own (partly to do with energy-positivity requirements), and it is my own view that such a perspective on an 'explanation for Λ' is not helpful. We shall be seeing later, however, that the presence of a positive cosmological constant is, in fact, *essential* to the particular viewpoint that I shall be putting forward here.

3.2 QUANTUM GRAVITY?

Up until this point, I have been considering spacetime only in its capacity of being describable in entirely *classical* terms. Quantum theory played a role, via the Planck formula $E = h\nu$, but we would be considering the spacetime itself as being an entirely *classical* structure within which the quantum activity of small physical systems would be viewed as taking place. However, this can be only a temporary viewpoint, since general relativity is a *dynamical* spacetime theory in which the gravitational field, as described by spacetime curvature, is influenced by its matter sources, and where these sources, being physical material, are taken to be ultimately quantum mechanical in nature. Accordingly, the spacetime structure must, itself, take on some quantum-mechanical features. Einstein's $\mathbf{G} = -8G\pi\mathbf{T}$ (or $\mathbf{G} + \Lambda\mathbf{g} = -8G\pi\mathbf{T}$) already tells us this, since the left-hand side cannot

[†] The results were obtained by two separate teams, headed respectively by Saul Perlmutter and Brian Schmidt; see Perlmutter (1998).

have an entirely classical description if the right-hand side is quantum mechanical.†

This is not to say that the quantum procedures that are to be applied to spacetime structure must necessarily be entirely conventional ones that are used in standard quantum field theories, aimed at providing a 'quantum-gravity theory' as that term is normally understood. Indeed, it is my strong opinion that the correct union of spacetime structure with quantum-mechanical principles will have to differ fundamentally from a standard quantum field theory. In Section 3.3, I shall describe some specific reasons for believing this, but aside from these, an overriding motivation for seeking a change in the structure of quantum theory is a belief that there are *paradoxes* in the application of strict quantum rules if they are to be applied indiscriminately at all levels. Basically, this is the contradiction between the probabilistic discontinuous procedure that is adopted in a 'measurement' in quantum theory and the deterministic continuous Schrödinger evolution of the entire quantum state, considered as encompassing both the quantum system under examination and the classically described measuring apparatus. Although such a belief seems to be somewhat at odds with the 'conventional' attitude to quantum theory, at least I can take comfort in the fact that at least three of the founders of that subject, namely Einstein, Schrödinger and Dirac, were all also of the opinion that quantum mechanics is a 'provisional' theory!

Of course, even if accepted, this belief does not imply that such needed changes would necessarily appear at a level at which the *gravitational* field becomes significantly involved. However, there are several reasons for anticipating that these changes in quantum theory are indeed to be expected at a stage where the principles of general relativity and observational aspects of the Universe (via thermodynamics) begin to have serious relevance. Two of these will have particular

† The suggestion that the '**T**' on the right be replaced by an *expectation value* $\langle \mathbf{T} \rangle$ has been made at various times, but this does not lead to a plausible theory. The remarks of Feynman *et al.* (1995) on this issue are pertinent here.

importance to this account, and I shall be coming to them shortly.[†] But for the moment, let us consider the *converse* issue of what the impact on *spacetime structure* may be expected to be, when the principles of quantum theory are taken into account.

One of the commonly accepted reasons for investigating the implications of 'quantum gravity' is that the classical theory of General Relativity runs into *singularities* in certain situations. These are situations where too much mass is concentrated into too small a region, such as in the collapse of over-massive stars or collections of stars to form black holes, and at the Big-Bang origin of the Universe. There are good observational (and theoretical) reasons to believe that these situations occur as actual features of the Universe we inhabit. It should be noted that the singularities that arise in the classical solutions of Einstein's equations are not mere features of the special symmetries possessed by the original exact models (Schwarzschild, and Friedmann–LeMaître–Robertson–Walker), but occur under *generic* circumstances, where no symmetry is postulated and only very weak causality requirements and local energy-non-negativity assumptions are made concerning the matter sources.[‡] An analogy has often been made with the pre-quantum-mechanical picture of a hydrogen atom where, in accordance with Maxwell–Lorentz theory, the electron should spiral into the proton nucleus with the emission of a burst of electromagnetic radiation, to reach a *singular* configuration at which the electron falls into the proton. Here, the coming of quantum mechanics resolved the issue, with a stable minimum-energy orbit which allowed no further emission of radiation. It is commonly assumed that the appropriate quantum-gravity theory ought to do something similar for the singularities in

[†] In Penrose (1996), (2000) and (2004, §§30.10-13 and note 30.37), I have, in addition, argued for the relevance of the principle of general covariance and the principle of equivalence; see also Diósi (1984, 1989) for earlier closely related proposals. It seems likely that such proposals might well be subject to experimental test before too long; see Marshall *et al.* (2003).

[‡] See Hawking and Penrose (1970).

black holes (and in the Big Bang[†]), but there is certainly no clear evidence for this, despite various claims that theorists have made from time to time. In fact, there are issues that come up here which render it most unlikely that such a straightforward-looking picture is likely to hold true, and these will play an important part in our later deliberations.

Another reason that has frequently been put forward for the necessity of having a consistent and believable quantum-gravity theory is that it ought to play a key role in resolving the ultra-violet divergence problems of quantum field theory. The basic problem is that in physical interactions, according to quantum field theory, one must sum up all the different processes that can take place, where some of these involve exchanges of particles of unboundedly large momentum. These summations (actually integrals) are usually *divergent*; the divergence (in the ultra-violet case, which is the most troublesome) is at the large-momentum end of the scale which leads to infinite answers. In quantum mechanics, large momenta correspond to small distances, so it is often argued that if there is some sort of 'cut off' at tiny distances these divergences could be replaced by finite expressions. The hope is that quantum gravity might achieve this in a natural way, the cutoff being at the 'Planck scale' of $\sim 10^{-35}$ metres.

The reason for believing that spacetime might begin to have a fundamentally different, and possibly *discrete*, character at this scale comes from the idea that 'quantum fluctuations in the gravitational field' – where this field is usually thought of as being described by a *quantised metric* **g**, according to the descriptions that most physicists seem to be comfortable with – and these fluctuations would destroy the appearance of a normal continuous manifold. The 'Planck' scale at which these fluctuations are considered to appear can be obtained simply from dimensional considerations: we take the fundamental

[†] The capitalisation of 'Big Bang' that I am adopting here is to distinguish the unique event that appears to have originated the Universe that we are familiar with from other hypothetical initial-type spacetime singularities.

FIG. 3.3 It has been argued that, owing to quantum fluctuations, at the scale of the Planck length of 10^{-35} m and Planck time of 10^{-43} s, spacetime ought to acquire a 'foam-like' structure.

constants c, G and $\hbar (= h/2\pi)$ and construct a distance – the *Planck length* – out of them:

$$\sqrt{\frac{G\hbar}{c^3}} \approx 1.6 \times 10^{-35}\ \mathrm{m}.$$

Or, in accordance with the notions referred to in Section 3.1, we might well consider the Planck *time* as being more fundamental

$$\sqrt{\frac{G\hbar}{c^5}} \approx 5.3 \times 10^{-44}\ \mathrm{s}.$$

John Wheeler has used the picture of a 'quantum foam' (see Figure 3.3) to illustrate one viewpoint as to what 'spacetime' might be like at the Planck scale. Quantum fluctuations are a consequence of Heisenberg's uncertainty principle, as applied to noncommuting pairs of field components. While it is certainly true that a classical description of a physical field will not do at a distance at which the scale of the fluctuations is comparable with the scale of the field itself, there is a tendency for physicists to form a rather classical-looking 'picture' of these things in which the field simply fluctuates wildly (as in Wheeler's 'foam' picture). A better picture might be in terms of some kind of 'noncommutative geometry'.[†]

[†] See the chapter by Connes in this volume and Connes (1994), as well as the chapter by Majid in this volume.

Many suggestions have been put forward as to the nature of a spacetime which might perhaps be *discrete* at its smallest scales. For example, it has been suggested, from time to time, that some kind of periodic lattice structure might provide an appropriate model.† Of course, such models encounter difficulties with the observed isotropy of space and with the approximate modelling of a curved macroscopic spacetime. Accordingly, other suggestions have been made in which the regular lattice is replaced by something with a more random-looking organisation, such as a *causal set*,‡ in which spacetime is taken to be modelled by a discrete system of 'events' for which it is merely specified which events causally precede which other events (mirroring the normal spacetime relation of which could be connected to which others by future-directed timelike or null curves). Again, it is quite plausible that some sort of noncommutative geometry might be of relevance here.§

Another suggestion of a different character is that provided by the theory of *spin-networks*. In its original form,¶ the idea was to regard spatial (or spacetime) points as being not basic at all, but to try to base things on *quantum angular momentum* (which is already something discrete in standard quantum mechanics). The model consists of basic elements referred to as *units*, each of which is represented schematically as a line segment (thought of as, in some sense, the 'history' of the unit) and is labelled by a non-negative *integer*, this integer being interpreted as twice the total angular momentum of the unit (i.e. twice its 'j-value'); but the unit is *not* assigned a value for its angular momentum about any given spatial direction (i.e. 'm-value'). The standard quantum rules whereby *probabilities* arise when units are combined or split apart in complicated combinations of ways are

† E.g. Snyder (1947), Schild (1949).

‡ E.g. Kronheimer and Penrose (1967), Sorkin (1991), Markopoulou (1998).

§ I am not aware that noncommutative in the sense of Connes has really been applied to 'spacetime' rather than just to 'space' or to other (internal) degrees of freedom, or if any connection can be made with the ideas of causal sets.

¶ The theory was originated in the mid 1950s, but not published (and even then not in full detail) until Penrose (1971).

then obtained, formulated in entirely combinatorial terms. Moreover, it can be shown that for large such 'spin-networks' these probabilities can be used to define a notion of 'angle' between the 'spin-axes' of the units which, in an appropriate limit, agrees with the way that angles behave in ordinary Euclidean space.

Generalisations of spin-networks have found their way into two different approaches to the combination of spacetime with quantum-mechanical principles. In the first place, attempts to transform spin-network theory into a 4-dimensional fully relativistic scheme formed one of the motivations behind *twistor theory*.[†] Whereas spin-network theory was developed from the representation theory of the spin version $SU(2)$ of the ordinary rotation group $SO(3)$, twistor theory arose partly from a study of the representation theory of the spin version $SU(2, 2)$ of the *conformal group* $SO(2, 4)$ of (compactified) Minkowski spacetime. The spacetime conformal group can be thought of as a group preserving the *light-cone* structure of flat spacetime, so it has some relation to the preservation of local spacetime causality.

The drive towards an analogue of a spin-network theory, in this context, may be regarded as one of the underlying motivations of *twistor diagram* theory[‡] which is the twistor version of the theory of Feynman diagrams of quantum field theory, and which has encountered something of a revival in recent years owing to some new input from string-theoretic ideas.[§] Twistor theory does not claim to be an approach to a quantum-gravity theory, as such, but provides a different outlook on spacetime, where nonlocal features arise at a fundamental level, and where the notion of 'spacetime point' (or 'event') is regarded as a secondary construct, the *light ray* (or, more correctly, the history of a massless entity with spin) provides something closer to the primitive element of the theory. Many aspects of physics, both classical and quantum, find neat descriptions in terms of complex-manifold

[†] See Penrose (1967, 1975, 1976, 2004, Chapter 33), Penrose and MacCallum (1972).

[‡] See Penrose and MacCallum (1972), Hodges (1982, 1985, 2006).

[§] See Witten (2004).

geometry in twistor theory, but the scheme is not completely wedded to the conventional formalism of quantum mechanics, there being some scope for generalisation to a *non-linear* scheme in which quantum state reduction might find some dynamical description.

Spin-network theory has also had some influence in the development of the *loop-variable* approach to quantum gravity.[†] Here, the basic quantum states are taken to be represented by spin-networks (slightly generalised from those that I had originally studied), but where rather than being the purely abstract combinatorial structures that I had initially had in mind, the spin-networks are now thought of as being imbedded in a continuous 3-manifold representing 3-dimensional space. Although formally these structures are indeed imbedded spin-networks (carrying some additional information from their topological imbedding), they have a rather different physical interpretation from the original spin-networks. Rather than representing *spin*, the numbers attached to the line segments actually represent an *area* measure, entirely concentrated along the line, in a delta-function manner. Spin-network lines that close into loops are indeed the loops which provide the 'loop variables' of the theory. This is a serious approach to a quantum-gravity theory which directly tackles some of the fundamental issues (such as general covariance or the initial basis states), initiated by Abhay Ashtekar and subsequently developed by many others.

In common with other approaches to quantum gravity, however, the loop-variable approach makes no attempt to break free of the standard procedures of quantum theory. As stated above, it is my own strong opinion that such a break will be necessary for a successful theory which is in accord with Nature. An important input into this belief comes from the above desire for a quantum gravity which resolves the spacetime singularity issue. And whatever theory this 'quantum gravity' might turn out to be, it ought to provide an

[†] See Ashtekar (1986), Ashtekar and Lewandowski (2004); also Penrose (2004, Chapter 32).

explanation for the *particular* structures that apply to the spacetime singularities of our actual Universe, these having an importance to physics that turns out to be quite fundamental, going far beyond the issues normally thought to lie within the scope of quantum gravity.

3.3 INPUT FROM THERMODYNAMICS

What are these structures, and why are they so fundamental to physics? Basically, the answer to the second question is that they provide the underpinnings of the *Second Law of thermodynamics*, and the specific *way* in which they do provides the answer to the first question. The Second Law tells us that a measure referred to as *entropy* has a relentless tendency to increase with time. How is 'entropy' defined? Roughly speaking, it is a measure of randomness of the state of a physical system but, more precisely, we can use Boltzmann's definition

$$S = k \log V$$

where S is the entropy and k is Boltzmann's constant (a small number having the value of about 1.38×10^{-23} joules per degree kelvin). To define the quantity V, we think in terms of the *phase space*† $\mathcal{P}$ of the system under consideration. The particular state of the system will be represented as a point q in $\mathcal{P}$. But as far as macroscopic parameters (such as pressure, temperature, fluid flow magnitude and direction, chemical composition, etc. but *not* fully detailed matters such as how some particular molecule might be moving, etc.) are concerned, there will be many other possible states of the system which are completely indistinguishable from q by any practical means. The collection of such points in $\mathcal{P}$ will have some (large-dimensional) volume, and this volume is V.

Of course, there is something a bit vague about this definition, because the notion of a 'macroscopic parameter' is a somewhat

† $\mathcal{P}$ has one dimension for each positional degree of the system and another dimension for each momentum degree of freedom, so for a system whose constituents are n structureless point particles, $\mathcal{P}$ has $6n$ dimensions; see Arnol'd (1978) for a more complete description, also Penrose (2004, §12.1, §§20.1,4, §§27.3-7).

imprecise and technology-dependent notion. One can well imagine that some future technology might be able to keep track of the behaviour of individual constituents of a system to a much greater extent than is possible today, and accordingly the volume V that is assigned to the system represented by q might then be somewhat smaller than would be estimated now. Nevertheless, in practice, the entropy concept is normally very robust, and issues of this kind can usually be safely ignored. The reason for this is that the different 'coarse-graining' regions (as they are called) that are distinguished by values of the macroscopic parameters tend to have volumes that are stupendously large and are vastly different from one another, and it is the combined effect of the logarithm in Boltzmann's formula and the smallness of Boltzmann's constant that has the effect that entropy values nevertheless turn out to be reasonable numbers and, moreover, that the entropy that is assigned to a particular state changes only very slightly when such refinements are introduced.

We can get a good feeling for why the entropy of a system has this relentless tendency to increase with time, when we bear in mind the vast differences in the sizes of the different coarse-graining regions. Imagine that our point q starts, at time t_0, in some reasonable-sized such region $\mathcal{Q}_0$, whose volume V_0 defines the entropy of the starting state. As the evolution of the system takes place, the point q wiggles around in a way that tends to move it from one coarse-graining region to another, the volume increasing by an enormous factor as time progresses, so that once it has found itself in such a region of a particular size, one can more-or-less ignore the possibility of finding a smaller one in the course of increasing time. In this way we can gain some understanding of the relentless entropy increase that the Second Law predicts.

However, this is only 'half' of the content of the Second Law, and it is overwhelmingly the 'easy' half. For it does not explain why, when we examine things in the *past* direction of time, the entropy actually goes *down* for times proceeding into the past from the given one t_0. When we work towards the past, we find that this kind of general reasoning simply does not work. The initial state of the Universe – the Big Bang – seems to have simply been *presented* to us as a state of extraordinarily small entropy and, because of the logarithm in Boltzmann's formula, an even *more* extraordinarily small volume $\mathcal{O}$ in the phase space $\mathcal{U}$ for the entire Universe.† We shall be seeing just *how* extraordinarily small $\mathcal{O}$ is shortly.

At this point, we should remind ourselves of one of the most powerful pieces of observational evidence for the actual occurrence of the Big Bang. This is the microwave background radiation, that comes to us from all directions and is frequently referred to as the 'flash of the Big Bang', cooled by the expansion of the Universe from an enormously high initial temperature to what we find at present, which is only about 2.7 K. In fact, the radiation that we observe comes not exactly from the initial flash, but from the state of affairs around 300 000 years after that event, when the Universe became transparent to electromagnetic radiation. The spectrum of this radiation is shown in Figure 3.4, indicating the intensity of the radiation for each frequency. We note that it is extraordinarily close to a *Planck spectrum* (the solid line in the diagram) which, bearing in mind the 500-fold exaggeration of the error bars depicted, gives an agreement between the observations and the theoretical Planck curve which, to the eye, looks perfect!

There is a seeming paradox in that the Planck spectrum is a clear indication that we are looking at a *thermal state* (i.e. the 'thermal equilibrium' that in a normal physical situation represents easily the largest coarse-graining volume in phase space), which refers to *maximum entropy* despite the fact that, according to the Second Law, the beginning of the Universe (as represented by the region $\mathcal{O}$ in $\mathcal{U}$) ought to be at a *minimum* entropy! I suspect that this seeming paradox worried cosmologists less than it should have because of some

† I hope that those readers who find themselves to be uncomfortable with assigning a phase space to the entire Universe may be somewhat reassured by the arguments given in §27.6 of Penrose (2004).

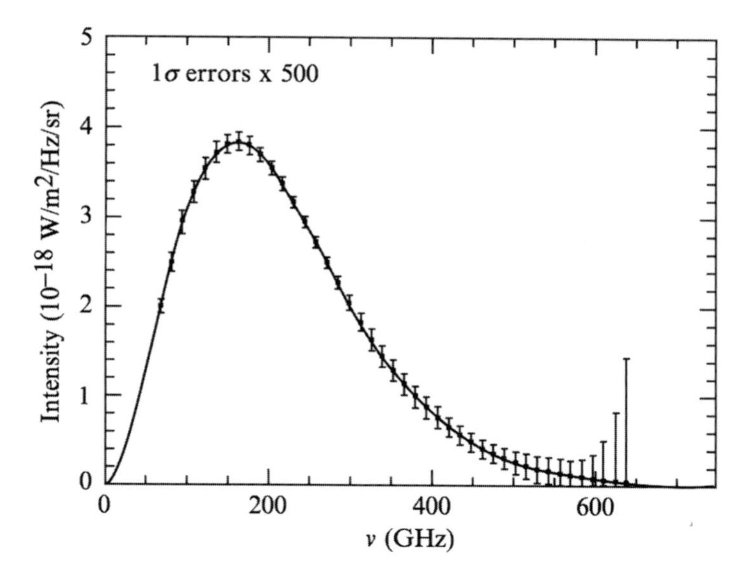

FIG. 3.4 The microwave background has an intensity in terms of frequency that is extraordinarily precisely in accord with the Planck spectrum, indicating thermal equilibrium for matter and radiation in the early Universe. (N.B.: error bars exaggerated by factor of 500.)

vague feeling that the Universe was very 'small' at the beginning, so that there would have been, in some sense, a 'low ceiling' on possible entropies. However, with the time-symmetrical physical laws – including the general relativity that governs the overall geometry of the Universe – that are normally assumed, this is *incorrect*, as can be seen by various arguments, such as the fact that the total phase space $\mathcal{U}$ is not supposed to 'evolve' with time; it is supposed to be just *one* space, in terms of which the evolution of the Universe is to be described. More specifically, we can consider the time-reversed situation of a collapsing universe. All its different states must evolve to *something*, so the large phase-space volume of these different possibilities cannot just shrink away. We must indeed think of (almost all, at least) of these different evolutions as ending in singularity, so their time-reverses would be possible 'Big-Bang' beginnings, and there is indeed no low ceiling on the different possibilities.

The answer to this riddle is that the thermal nature of the microwave background is telling us merely that what we are looking at is *radiation* that was in thermal equilibrium with *matter*; it is telling us nothing about the gravitational field. What we also see in this radiation is that it is very nearly uniform over the entire sky (to about one part in 10^5). This is a strong indication that the Universe, at the time of the Big Bang was very closely isotropic about our spatial location, and if we assume that there is nothing special about our spatial location, then the Universe must be closely spatially homogeneous also. This spatial isotropy and homogeneity is an indication that the *gravitational* degrees of freedom were not excited at the beginning of the Universe – and certainly not 'thermalised'. What we find is that the extreme lowness of the entropy of the Big Bang can be attributed to this fact.

The initial uniformity of the material of the Universe provided the opportunity for gravitational clumping, and the lowness of the entropy of the early Universe lies in the potential for this clumping. This gravitational clumping is relentless and ultimately gives an enormous entropy increase when material collapses to black holes. At the intermediate stage of our present experiences, we take advantage of the fact that the Sun is a hot spot in a dark background sky. The plants on this Earth make use of this imbalance through the process of photosynthesis, and the present cycle of life on Earth depends upon it.

The formation of black holes represents a vast increase in the entropy, and by use of the Bekenstein–Hawking black-hole-entropy formula, we can estimate, roughly, the entropy that is available in, say, the baryonic matter within our observable Universe. We do this by imagining that this entire matter were to collapse into a black hole. This entropy comes to around 10^{123} and we can use this figure to obtain a roughly estimated value for the size of the total phase space $\mathcal{U}$ (although we would get a somewhat larger value if we were to include the *dark matter*). This gives a total phase-space volume of around $10^{10^{123}}$ (or $e^{10^{123}}$; it makes no significant difference). The entropy in the microwave background (which, after its discovery, had been regarded as being extraordinarily large, swamping all the then known other physical contributions) within the observable Universe is about 10^{88}, giving a phase-space volume of $10^{10^{88}}$. Owing to the significant number of large black holes in the observable Universe (in galactic centres), the present entropy value is probably more like $10^{10^{101}}$. In any case, these numbers are completely swamped by the $10^{10^{123}}$ value, and we find that the precision in the Big Bang (just with regard to the observable part of our Universe, but with the precision presumably extending far beyond this) is about one part in at least $10^{10^{123}}$, although the presence of dark matter of about 10 times the mass of baryonic matter could put the phase-space volume up to around $10^{10^{125}}$.

We see, in such figures, the absolutely stupendous precision that must have been present in the Big-Bang singularity, this precision being something that apparently refers entirely to *gravitation*. On the other hand, the singularities in black holes cannot be constrained in this way, both for the reason that if they were then we should have a blatant violation of the Second Law, and also for spacetime-geometrical reasons of conflict between the internal geometry of black holes and that of homogeneous isotropic Universe models. Whatever quantum-gravity scheme is put forward in order to explain what goes on at the regions which are classically referred to as 'spacetime singularities', it must, it seems to me, account for this manifest time-asymmetry. Classical General Relativity is a completely time-symmetric theory, and so are the standard procedures of quantum (field) theory. I do not see any way of providing the needed gross time asymmetry without breaking free of this conventional mould of 'quantum-gravity' procedures.

To end this section, I need to refer to one further related issue, which will have fundamental importance to the ideas that I am putting forward in Sections 3.4–3.6. It may be reasonably argued that the one clear-cut application of the bringing together of procedures from quantum mechanics and general relativity is in the Bekenstein–Hawking black-hole entropy and the accompanying Hawking temperature of a black hole. But one of the implications of the black-hole temperature is that although it is extremely cold for any astrophysical black hole (about 10^{-7} K, for a black hole of stellar mass, and far colder still for a black hole at the centre of a typical galaxy), the accelerating expansion of the Universe is expected to result in ambient temperatures that are lower even than these values. After that, the black holes will gradually lose energy by Hawking evaporation until after some enormous period of time they are expected to lose all their mass and, with increasing temperature, probably evaporate away completely, in a final 'pop'.

An issue that has puzzled theorists is what happens to the 'information' that went into their formation, since the Hawking radiation is supposed to be completely thermal and therefore 'information-free'. This is sometimes referred to as the *information paradox*. A consensus had seemed to have been building up in recent years (including a change of mind on the part of Hawking himself) to the effect that this 'information' is somehow all emitted back by the hole, perhaps hidden in subtle correlations in the radiation.[†] From the point of view of the usual geometry of a black hole, this seems to me to be exceedingly unlikely – particularly in view of the fact that the original Hawking derivation[‡] of the black-hole temperature (the only one which, to my mind, instils clear-cut physical conviction) depends essentially on this 'information loss'). The basic presumption that somehow such information *cannot* be lost comes from a belief in the fundamental inviolability of quantum theory's bedrock: *unitary evolution* (i.e. evolution according to the Schrödinger equation). In my own view, this is a poor reason, as unitary evolution is in any case violated in the standard procedures that are adopted in the measurement process (see Section 3.2). In fact, I would regard the 'information paradox' and the

[†] See, for example Bekenstein (1993), Hawking (2005); but for a counter argument, see Braunstein and Pati (2007).

[‡] Hawking (1975).

'measurement paradox' as closely linked, where both point to a need for a fundamental change in the formulation of quantum mechanics. I shall come back to this issue later.

3.4 CONFORMAL BOUNDARIES TO SPACETIME

Let us return to the very special structure of the Big Bang, and try to characterise this geometrically. What seems to be required is some geometrical formulation that describes the suppression of *gravitational* degrees of freedom in the initial state. Let us first consider the analogy of Maxwell's electromagnetic theory. Here, there are two tensor fields that play basic roles, namely the $\left[{0 \atop 2}\right]$-valent *Maxwell field tensor* **F**, encompassing, in a relativistic way, both the electric and magnetic fields, and the $\left[{0 \atop 1}\right]$-valent *charge-current vector* **J**, encompassing the electric charge density and the electric current vector density. We think of **J** as representing the *source* of the electromagnetic field **F**. In Einstein's theory, the source of gravitation is taken to be the $\left[{0 \atop 2}\right]$-valent trace-reversed Ricci tensor **G**, as we have seen. Although it is not so frequently expressed in this way, there is also a clear-cut analogue of Maxwell's **F** in gravitation theory, and this is the *Weyl conformal tensor*, which is a $\left[{0 \atop 4}\right]$-valent tensor **C** obtained by appropriately removing the information contained in the Ricci tensor **P** from the Riemann–Christoffel tensor **R** (where **C** is obtained from **R** by subtracting a certain expression constructed from **P**, **g** and $\mathbf{g}^{-1}$, so as to obtain a completely trace-free quantity).[†]

Weyl's tensor **C** has a direct physical/geometrical interpretation in terms of its effect on light rays. Whereas the Ricci tensor **P**, in physical situations, acts as a *magnifying* lens (actually it is the trace-free part of **P** that achieves this; the trace itself has no visible effect on light rays), the Weyl tensor **C** has a purely *astigmatic* influence on light rays, distorting what would otherwise be small circular patterns into

[†] See, for example, Penrose (2004, end of §19.7). Often the Weyl tensor is presented in a $\left[{1 \atop 3}\right]$-valent form (which I am calling $\mathbf{C}^{\hat{}}$), obtained by 'raising one of its indices' with $\mathbf{g}^{-1}$.

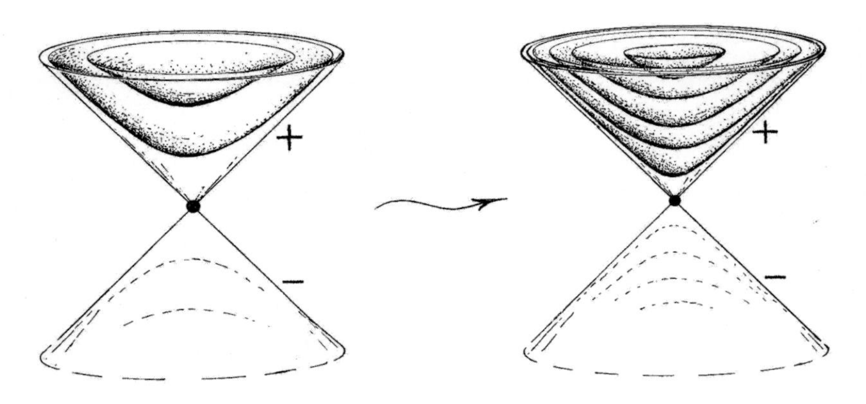

FIG. 3.5 A conformal rescaling preserves the (±) light-cone (i.e. causal) structure of spacetime, but clock rates are altered. The hyperboloidal surfaces represent equal time separations from the cone's vertex. The degree of crowding of these surfaces (clock rates) is what is altered by a conformal rescaling.

elliptical ones. This distortion is a now familiar effect of *gravitational lensing*.

The role of **C** as a measure of conformal curvature arises when we consider *conformal rescalings* of the spacetime metric:

$$\mathbf{g} \mapsto \Omega^2 \mathbf{g}.$$

Here, Ω is a (normally positive-valued) scalar field. Spacetime conformal rescalings are changes in the spacetime metric which leave the light cones unaffected; see Figure 3.5. We can consider that the labelling of the two half light cones as 'past' and 'future' is also left unchanged; accordingly, we can think of conformal rescalings as precisely the changes in **g** that leave the *causal relations* between points unaffected.

Conformal rescalings are also frequently used in the Riemannian (positive definite) case. A good example of this is in the Beltrami–Poincaré representation of a 2-dimensional space known as the *hyperbolic plane* (a form of non-Euclidean geometry), beautifully illustrated in the print 'Circle Limit IV' by M. C. Escher, reproduced here as Figure 3.6. We are to imagine that in the Riemannian geometry being represented, the angels are all the same size and shape as each other,

FIG. 3.6 Escher's '*Circle limit IV*'† illustrates how the entire Riemannian geometry of the non-Euclidean (hyperbolic) plane can, by use of conformal rescalings, be accurately represented as the interior of a Euclidean disc. Note that the 'infinity' of the hyperbolic plane is represented conformally as a *smooth boundary* to the geometry, in a way that anticipates the procedures described in Section 3.4 for representing infinity (or certain singularities) as conformal boundaries.

and so also are the devils. In the Euclidean representation, which is the interior of a circular disc, they look smaller and more crowded towards the edge of the disc. But if we expand them out by a conformal rescaling (which preserves angles and infinitesimally small shapes), we find that the conformal geometry of the figures near the edge is identical with that near the centre. Thus, the conformal geometry of the interior of a Euclidean circular disc is identical to that of the entire hyperbolic geometry. This figure also illustrates another feature, which will be central to the discussion later in this section: we find that the *infinity* of the hyperbolic geometry can be represented conformally as a smooth boundary, here the circle bounding the disc.

It should be noted that, of the 10 independent components that **g** has altogether at any one point in spacetime, 9 of them simply determine where the light cone is at that point (these being the 9 independent *ratios* of the 10 independent g_{ij}). The remaining independent component, in effect, provides the *scale* of the metric at that point. Since the spacetime metric is really a measure of clock rates, this scale is what fixes the actual passage of time, once the causal structure of the spacetime has been determined. In a clear sense the causal structure is more fundamental than the time rates. There are various parts of physics which depend only on this causal (or conformal) aspect of **g** and actually do not depend on this scaling. The Maxwell theory of electromagnetism, which is conformally invariant,† is a good example. Basically, it is *mass* and *gravitational interactions* that need the scale (and we shall be examining this issue a little more carefully later). Photons are massless entities and are insensitive to this scaling.

Under the above conformal rescaling, we find that **C** is a *conformal density*, in the sense

$$\mathbf{C} \mapsto \Omega^2 \mathbf{C}.$$

It may be noted that if we 'raise an index' with $\mathbf{g}^{-1}$ to get a $\begin{bmatrix} 1 \\ 3 \end{bmatrix}$-valent version $\mathbf{C}^{\hat{}}$ of the Weyl tensor, we get invariance without the Ω^2 factor. If we raised *two* indices (to get $\mathbf{C}^{\hat{}\hat{}}$), however, we would find a factor Ω^{-2} on the right, etc. It is, accordingly, appropriate to think in terms of conformal densities like these, rather than to concentrate only on quantities (like $\mathbf{C}^{\hat{}}$) that are conformally fully 'invariant', in the sense that they transform without a power of Ω. Accordingly, I shall henceforth use the term 'conformal invariance' in the more general sense which includes conformal densities of this kind.

On the other hand, **P** and **G** transform in much more complicated ways, and are certainly not conformal densities. More

† See, for example, Penrose and Rindler (1984).

significantly, $\mathbf{C}$ provides the measure of *conformal curvature*, i.e. the part of the spacetime curvature which measures the causal structure (locally). The condition for an open region of spacetime to be conformally flat (i.e. having the same causal structure, locally, as Minkowski space $\mathbb{M}$) is $\mathbf{C} = 0$.

Let us now try to characterise the kind of restriction that seems to have held at the Big-Bang singularity of our actual Universe. To say that the gravitational degrees of freedom were somehow not activated at that time would be to assert that something of the nature of '$\mathbf{C} = 0$ at the singularity', or, at least, that $\mathbf{C}$ should be severely constrained there. Indeed, this condition is one that I proposed in around 1978 as the 'Weyl curvature hypothesis',[†] this being a grossly time-asymmetric restriction on past-type spacetime singularities, which is supposed not to apply to the future-type singularities that appear in black holes. However, as stated simply in this way, there is some mathematical awkwardness about this proposal. There are different interpretations as to what it might mean to say '$\mathbf{C} = 0$', or '$\mathbf{C}$ is bounded', or '$\mathbf{C}$ is small' in some other sense, at a *singularity*, where the very manifold structure, let alone the metric, would be expected to be undefined (not to mention how 'quantum fluctuations' in the curvature field at the Big Bang might be perceived as affecting the issue). One can talk about *limits*, like '$\mathbf{C} \to 0$' or '$\mathbf{C}$ remains bounded' as the singularity is approached, but this also may lend itself to numerous different possible interpretations, especially as the precise statements may vary, depending on which of the possible versions $\mathbf{C}$, $\mathbf{C}^{\wedge}$, $\mathbf{C}^{\wedge\wedge}$, $\mathbf{C}^{\wedge\wedge\wedge}$ or $\mathbf{C}^{\wedge\wedge\wedge\wedge}$ one might choose to use, these being defined by having indices raised successively.[‡]

In view of such uncertainties, it is gratifying that there is another way of stating things, which is considerably more mathematically satisfying and relatively free of ambiguities. This is an approach which has been clearly formulated and extensively studied by my Oxford

[†] See Penrose (1979).

[‡] Ambiguities arising from different index orderings are not significant here.

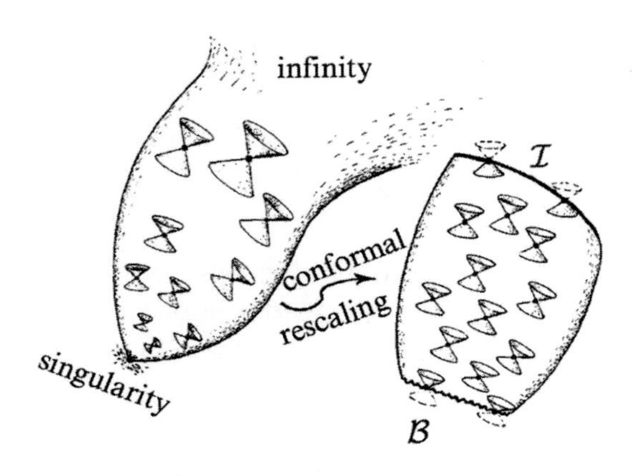

FIG. 3.7 In favourable circumstances, the conformal structure of a pseudo-Riemannian manifold can be extended to regions where the (pseudo-)metric is not defined. This procedure can be used to adjoin smooth boundaries and can be used to 'bring into view' the structure of the Big Bang, or of infinity.

colleague Paul Tod.† This takes advantage of an idea that has been used in General Relativity theory for over 40 years whereby, in suitable circumstances, a cosmological singularity can be represented as a *conformal boundary* to a spacetime (where an infinite conformal expansion is involved) and so also can the asymptotic (infinite) region be so represented (involving an infinite conformal contraction); see Figure 3.7. Indeed, to express the required smoothness conditions succinctly, one can envisage extending the spacetime, as a conformal manifold *beyond* such a boundary. Thus, rather than specifically referring to the Weyl tensor, one phrases the required condition on the Big Bang in terms of conformal spacetime geometry, as the condition that the conformal geometry of the spacetime manifold $\mathcal{M}$ can be *extended in a smooth way* to a somewhat larger 'conformal spacetime' $\mathcal{M}^*$ that includes an open region $\mathcal{M}^-$ *preceding* the Big Bang. The Big Bang itself is represented as a (spacelike) smooth 3-manifold

† See Tod (2003). For the use of such boundaries in cosmology generally, see Penrose (1964), Penrose (1965), Penrose and Rindler (1986).

$\mathcal{B}$, which acts as a past boundary to $\mathcal{M}$ and also as a future boundary to the attached region $\mathcal{M}^-$, the two regions $\mathcal{M}$ and $\mathcal{M}^-$ joining smoothly along $\mathcal{B}$, to form the combined manifold $\mathcal{M}^*$. This involves a conformal rescaling of the metric of $\mathcal{M}$, 'stretching' it out with a conformal factor Ω which becomes *infinite* at $\mathcal{B}$. In fact, with plausible-looking physical assumptions as to the relativistic equations of state for the initial matter distribution, we find that we can arrange that Ω^{-1} goes smoothly to 0 at $\mathcal{B}$ at the same sort of rate as a cosmological time parameter t, and we may schematically write this as

$$\Omega \sim t^{-1}.$$

No physical reality is attached to this 'pre-Big-Bang' phase $\mathcal{M}^-$; it is introduced solely as a 'mathematical device', so that the required condition on the physical region $\mathcal{M}$ can be neatly formulated.

It is an implication of this conformal extendibility that the Weyl curvature indeed turns out to be severely constrained. We see immediately from the assumed conformal regularity of $\mathcal{M}^*$, and the consequent finiteness of the Weyl tensor that $\mathcal{M}^*$ has at $\mathcal{B}$, that the Weyl tensor $\mathbf{C}$ of $\mathcal{M}$ (with its original physical metric) must satisfy

$$\mathbf{C} \sim t,$$

which describes a rate at which $\mathbf{C}$ approaches zero at $\mathcal{B}$. However, the status of a 'zero limiting value' of this tensor quantity is somewhat unclear, especially when we take into account the fact that if we had used $\mathbf{C}^{\hat{}}$ in place of $\mathbf{C}$, we would find merely the implication that the $\mathbf{C}^{\hat{}}$ of $\mathcal{M}$ (with the physical metric) must remain bounded as the Big Bang is approached, and had we used $\mathbf{C}^{\hat{}\hat{}}$, then merely $\mathbf{C}^{\hat{}\hat{}} \sim t^{-1}$. (In fact, in one sense, $\mathbf{C}^{\hat{}\hat{}}$ is the most reasonable measure to use since it is the least sensitive to 'coordinate dependence' of these tensorial measures since $\mathbf{C}^{\hat{}\hat{}}$ can be used to define scalar invariants without the further use of $\mathbf{g}$ or $\mathbf{g}^{-1}$.)

It will turn out, however, that the proposal that I am putting forward in Section 3.5 has the clear implication that a *stronger* condition holds at the Big Bang, namely that the Weyl tensor of the conformally rescaled space actually *vanishes* at $\mathcal{B}$. We then find that for the physical metric, as the Big Bang is approached,

$$\mathbf{C} \sim t^2, \quad \mathbf{C}^{\hat{}} \sim t, \quad \text{and} \quad \mathbf{C}^{\hat{}\hat{}} \quad \text{remains bounded.}$$

This is perhaps more in line with the intentions of the Weyl curvature hypothesis, as described in Section 3.3.

Let us now turn our attention to the 'other end' of the Universe, namely the remote future. Taking present-day observations at their face value, and adopting Einstein's $\mathbf{G} + \Lambda\mathbf{g} = -8G\pi\mathbf{T}$ as providing a correct description of the ultimate fate of the Universe, we seem driven to the viewpoint that Λ has a small (but cosmologically significant) positive constant value, implying that the Universe will continue to expand indefinitely, and at an accelerating rate. We may regard the de Sitter model, introduced by Willem de Sitter in around 1917, as a reasonably accurate description of the remote future of the spacetime geometry of the actual Universe that we inhabit. In de Sitter's model, the energy-momentum tensor $\mathbf{T}$ is taken to be zero, but that is not so unreasonable, for a first approximation to the situation of the very remote future, as the density would indeed be expected to go to zero as a result of the accelerated expansion.

In the case of de Sitter space – and, in fact, for a broad class of expanding-universe models, with positive Λ – we can use a kind of reverse mathematical trick to the one which Tod employed. By 'squashing the metric down' by a conformal rescaling factor Ω that tends to *zero* ($\sim t^{-1}$), the future infinity can be made into a finite future boundary $\mathcal{I}$ to the spacetime $\mathcal{M}$, which could be mathematically extended to beyond $\mathcal{I}$ in a smooth way, the conformal metric being completely regular across $\mathcal{I}$. This kind of trick has been used since the mid 1960s to study the properties of gravitational (and electromagnetic) radiation,† providing a powerful means of studying detailed behaviour (often in a geometrical way) without recourse to complicated asymptotic limits. It has been used most frequently in

† See Penrose (1965), Penrose and Rindler (1986).

the case $\Lambda = 0$, where $\mathcal{I}$ turns out to be a *null* (i.e. lightlike) hypersurface, but here there are complicating issues, still not fully resolved, about the generality of the approach. In the case $\Lambda > 0$ that concerns us here, the hypersurface $\mathcal{I}$ is necessarily *spacelike* and there are now powerful mathematical theorems demonstrating that the approach in fact has full generality in this case,† provided that (massive) matter sources are absent in the asymptotic future.

What about our actual observed Universe then? In the Universe's late stages, a major family of constituents will be numerous black holes. There will be some 'rogue' black holes each of a few solar masses (formed by stellar collapse), having escaped from their parent galaxies. But large black holes, situated at the centres of galaxies (such as the three-million solar mass black hole at the centre of our own Milky Way) are likely to form one of the main contributions to the total mass of the Universe in its very late stages. We must expect that, over time, a large fraction of the galaxy's mass (stars, smaller black holes, presumably also dark matter) will be consumed by such galactic black holes, but again some sizeable fraction of the galactic mass may be expected to escape as 'rogue' material. But in bound galactic clusters, galactic collisions will eventually become commonplace, with the congealing of black holes to form even larger ones settling at the centres of the clusters. As repeated collisions take place, involving larger and larger conglomerations of galaxies, a greater and greater proportion of the overall material of the original cluster (both baryonic and dark, and other black holes) will find its way into a central black hole of enormous mass.

As the Universe expands, its temperature reduces to below the Hawking temperature of even the largest of the black holes, although this will take a very long time. Eventually, therefore, Hawking evaporation takes over and after a far longer time scale (around 10^{67} years for a black hole formed from stellar collapse, and some 10^{94} years for a huge 10^{10} solar mass black hole) the holes would be expected to

† See Friedrich (1998).

disappear completely after radiating – primarily as *electromagnetic radiation* – their entire mass away. In the final 'pop', there would be the production of other kinds of particle (such as baryons, electrons and neutrinos) but mass-energy in this form would constitute only an extremely tiny part of the total.

Thus, we are left with a universe filled to a large extent with electromagnetic radiation (photons), but with a highly significant proportion of the energy density also in the form of gravitational radiation (gravitons). This latter contribution arises primarily from the in-spiralling of black holes that mutually swallow one another. Perhaps something of the order of 10% of the energy stored in the black holes would be emitted in this form in each black-hole collision, but since this is repeated many times in the process whereby the holes ultimately congeal into a single central black hole – for each bound galactic cluster or supercluster – we must expect that ultimately a fraction approaching 100% of its energy goes away into space in the form of gravitational waves, to accompany the electromagnetic energy emitted by the Big Bang, by stars, by the eventual Hawking radiation from evaporating black holes and by other thermal sources.

Now an important feature of electromagnetic radiation is that the equations that govern it are conformally invariant, so that this radiation is sensitive only to the conformal structure of spacetime. What about gravitational radiation? Here, what appears as a curious mathematical quirk of Einstein's theory comes into play. We may define a $\begin{bmatrix} 0 \\ 4 \end{bmatrix}$-valent tensor $\mathbf{K}$, taken to be exactly equal to the Weyl tensor $\mathbf{C}$, when we consider the original physical metric $\mathbf{g}$, but where we choose to scale it differently from $\mathbf{C}$ when the metric undergoes the conformal rescaling $\mathbf{g} \mapsto \Omega^2\mathbf{g}$, namely

$$\mathbf{K} \mapsto \Omega\mathbf{K},$$

in contrast to the $\mathbf{C} \mapsto \Omega^2\mathbf{C}$ that we had for the Weyl tensor proper. We find that with this scaling, the *free-field equations* for $\mathbf{K}$ share the conformal invariance that is a feature of the field equations of other massless fields, but in this case it is the equations for 'spin-2', the source-free Maxwell equations being the 'spin-1' case. (In the gravitational case, the free-field equations are what are called the 'Bianchi identities' in the situation when $\mathbf{P} = 0$.)† I shall refer to **K** as the 'graviton-field' tensor, and it is conformally invariant in the same way that the 'photon-field' tensor **F** is conformally invariant.

In fact, such conformal invariance, for some physical material, has the implication that it is not possible to build a 'clock' entirely of that material. For if such a clock could be built, the application of a conformal rescaling would not affect its physical operation, whereas the rescaling would certainly affect the physical time rate, as defined by the metric **g**. Hence the putative clock would actually be incapable of measuring that time rate!

In the case of electromagnetism, all its *interactions* with sources – i.e. the interactions between photons and charged particles – are also conformally invariant. But in the case of gravity, this is not so. However, because of the difference in scaling behaviour between **K** and **C**, the gravitational interactions effectively die away at $\mathcal{I}$ so, in effect, the gravitational field (as described by **K**), like the electromagnetic field (as described by **F**), becomes *ultimately* conformally invariant. Despite the material in the Universe becoming more and more attenuated as it becomes involved in the exponential expansion, we find that according to the metric scaling whereby $\mathcal{I}$ is a finite bounding hypersurface, both of the (rescaled) fields **F** and **K** remain *finite* at $\mathcal{I}$ and leave their non-trivial marks in the detailed structure there. This structure keeps record of all the electromagnetic and gravitational radiation that escapes absorption. Moreover, we note that because of the scaling differences between **K** and **C**, the (rescaled) Weyl curvature actually *vanishes* at $\mathcal{I}$. It is **K** (appropriately re-scaled on $\mathcal{I}$) that keeps track of the gravitational radiation – although we can see this also in the *normal derivative* of the Weyl curvature at $\mathcal{I}$ (and also in the intrinsic 3-space conformal curvature of $\mathcal{I}$).

† For a full discussion of these matters, see Penrose (1965), Penrose and Rindler (1984, 1986).

3.5 CONFORMAL CYCLIC COSMOLOGY

Let us now come to the main idea of *conformal cyclic cosmology* (CCC)†. In Section 3.4 we considered two complementary mathematical procedures which, up until this point, we considered merely as 'tricks'. One of these provides a formulation of the Weyl curvature hypothesis (WCH) by postulating that the conformal spacetime geometry extends smoothly to before the Big Bang across a conformally smooth hypersurface $\mathcal{B}$, to a hypothetical 'pre-Big-Bang' phase. The other provides us with a geometrical framework for describing the asymptotic future of the Universe in an elegant way, by allowing the conformal spacetime geometry of the Universe to extend smoothly to beyond the future infinity across a conformally smooth hypersurface $\mathcal{I}$. In neither case is the 'conformal spacetime' lying beyond the boundary ($\mathcal{B}$ and $\mathcal{I}$, respectively) assigned any physical reality. In each case, the boundary was introduced merely as a mathematical convenience. But here is where CCC comes in. It demands that *both* extended regions are now to be taken seriously as *real* universe regions!

The physical justification of this is, first, that in the *remote future* – as we have seen earlier – the major constituents of the Universe may well be conformally invariant (photons) or eventually *effectively* conformally invariant (gravitons), and there would ultimately be no way of constructing a 'clock' out of such material. Another way of addressing this issue is to think of how the 'passage of time' would be assessed by such massless entities as photons or gravitons. As was indicated in Section 3.1, according to relativity, a massless particle does not experience any passage of time. Using the Planck/Einstein relations referred to earlier, we find that 'eternity' (i.e. $\mathcal{I}$) occurs before the first 'tick' of the 'clock' that such a particle would define according to the prescription put forward in Section 3.1. Thus, these massless entities, being (effectively) conformally invariant, lose track of the rate of the passage of time, and would be more concerned with the

† See Penrose (2006).

conformal (or causal) geometry of $\mathcal{M}^*$, in the neighbourhood of $\mathcal{I}$, than with the metric geometry of $\mathcal{M}$, so we could adopt a viewpoint that the exact location of $\mathcal{I}$ within $\mathcal{M}^*$ would not be of particular concern to the eventual universe's contents or dynamics!

Now let us consider the remote *past*, as we approach the Big Bang backwards from the future. Near the Big Bang, we find that energies get so enormously large that mass becomes of no consequence, and the physics in the earliest stages – including all relevant interactions – again becomes *conformally invariant*. This assertion demands that all particle interactions become conformally invariant in the high-energy limit. This appears to be largely consistent with current particle-physics understanding. The standard treatment of particle interactions (strong, electro-weak) in particle physics is indeed in accordance with a scheme that becomes conformally invariant at high energies (basically when energies greatly exceed the 'Higgs mass'). Thus, near the Big Bang, there are, similarly to the situation in the remote future, no 'clocks', and it is again the *conformal* (or causal) structure of spacetime that is relevant to physical behaviour. One might say that, near the Big Bang, the Universe 'has no conception of the passage of time'[†]. Accordingly, the Universe, in a sense, is not sensitive to the actual 'moment' at which the Big Bang took place, and it is the conformal geometry, not the full metric geometry of spacetime, that is relevant there. Thus we may take the view that, as with the situation at the ultimate future hypersurface $\mathcal{I}$, it is the (conformal) geometry of $\mathcal{M}^*$, that is relevant, in the neighbourhood of $\mathcal{B}$ also, and the exact location of $\mathcal{B}$ within $\mathcal{M}^*$ is not of primary concern to the Universe's dynamics.

There is, however, a critical issue about *gravitational* interactions in the vicinity of the Big Bang that I have glossed over so far. The problem is that rather than dying away as the Big Bang is approached, the effect of gravity *increases*, at the higher and higher energies and densities that are expected to be encountered at the Big Bang. The

[†] See Rugh and Zinkernagel (2007).

key idea for resolving this issue is that, largely because of the form of WCH that is being adopted, gravitation can actually be ignored at the Big Bang (as a first approximation at least). This particular form of WCH demands that the Weyl curvature actually *vanishes* at the conformal hypersurface $\mathcal{B}$ which represents the Big Bang. How does this restriction come about? This is where CCC comes in.

The fundamental postulate of CCC is the (outrageous?) proposal that what lies 'beyond' the future boundary hypersurface $\mathcal{I}$ is the *big bang* of a 'new universe' and, correspondingly, what lay 'before' our Big-Bang hypersurface $\mathcal{B}$ was the *future infinity* of a 'previous universe'! The viewpoint is that in each case, the dynamics of the Universe could be carried across the bounding hypersurface ($\mathcal{I}$ and $\mathcal{B}$, respectively). This would be a *spacelike* hypersurface in each case – where, for $\mathcal{I}$, this depends upon the presence of a *positive cosmological constant* – so there is indeed the geometrical possibility of some sort of identification between the two. Rather than identifying the $\mathcal{B}$ and $\mathcal{I}$ for the same universe, however (which could lead to awkward problems with regard to causality), the CCC proposal is to postulate a succession of universes – or of universe *aeons* in the terminology that I am suggesting here – in which the $\mathcal{I}$ of one aeon is to be identified with the $\mathcal{B}$ of the next, in unending succession.

Many readers may find it puzzling how a particular aeon, when its expansion has continued out to infinity in the future, its density vanishing, and its temperature also reducing to nothing[†] can somehow be equated to an initial state with the seemingly opposite characteristics of infinite compression, infinite density and infinite temperature. It would indeed make no sense, geometrically or physically, if the *metric* geometries of the two regions were to match,

[†] There is a 'Hawking' temperature associated with the de Sitter space, of ultimate expansion (see Gibbons and Perry (1978), Gibbons and Hawking (1993)), far lower than the lowest Hawking temperature of any black hole that could exist on its own, which has been argued would be still present in the final state of the Universe. The arguments pointing to the presence of such a lowest temperature are, to my mind, somewhat questionable, but in any case such a lowest temperature does not affect any of the discussions presented here.

where this would indeed entail matching the physical quantities of mass density and temperature on one side with those corresponding quantities on the other. But mass density and temperature depend on the metric structure of spacetime, and cannot be determined from merely the conformal (i.e. causal) spacetime structure. In this proposed scheme, the idea is that in the remote future and in the remote past, the physical material of the Universe is sensitive only to the nine components per point provided by the *conformal* (light-cone) structure of the Universe and is blind to the remaining one component that provides the *scale* of the metric which would determine clock rates and hence distance measures. In terms of what the contents of the Universe can actually measure, the state of the Universe in the remote future is *identical* with the state of a big bang. At least, that is the idea; at the moment, it remains to be determined to what extent this very radical proposal is consistent with observational fact.

In this proposal a strong form of WCH holds, in which we have not only a smooth conformal extension to the past of $\mathcal{B}$, but one in which the Weyl curvature *vanishes* at $\mathcal{B}$. This comes about because of the identification being made between the $\mathcal{I}$-hypersurface of the previous aeon and the $\mathcal{B}$-hypersurface of the current aeon. Since the Weyl curvature actually vanishes at the previous aeon's $\mathcal{I}$, the Weyl curvature must also vanish at the $\mathcal{B}$ of the current aeon, the joined-up $\mathcal{M}^*$-space being assumed to be a *smooth* conformal 4-manifold across this joining hypersurface. This strong requirement seems to be what is needed for it to be consistent to ignore potentially infinite gravitational influences to enter the picture at the Big Bang which could provide fundamental difficulties for the current proposal (or, indeed, for most other proposals) for the treatment of the Big Bang. Even so, there are several as yet unresolved issues here, which have the current nature of 'work in progress'.

It would be natural to take the view that, whatever its merits or de-merits might be, such a fantastical scheme must surely be beyond any possibility of scientific scrutiny, and would lie in the province of philosophy or, perhaps, science fiction. However, it seems to me that there are many fairly clear-cut implications of CCC which ought to be of direct observational consequence. In the first place, there is the issue of whether the conditions at the Big Bang are indeed of the kind predicted by CCC, which implies a special form of WCH. This might, for example, have particular consequences with regard to the presence – or otherwise – of primordial gravitational radiation.[†] It would seem that the level of primordial gravitational radiation in the CCC proposal would be likely to be much too low to be detectable by the present (operational or under construction) ground-based gravitational wave detectors, or by the projected LISA space-based detector. On the other hand, the fashionable 'inflationary scenario' appears to predict an intensity of primordial gravitational radiation that might be measurable by LISA. Inflationary cosmology posits an *exponential expansion* for the Universe, between the first 10^{-35} and 10^{-32} seconds, or so, of its existence, and I shall refer to some of its implications in what follows.

Of particular relevance are the small variations (of the order of 1 part in 10^5) in the temperature of the microwave background that are observed over the celestial sphere. These temperature variations have at least two characteristics which have appeared to lend strong support to inflation. One of these is a *scale invariance* which is interpreted, in the inflationary picture, as being the result of the exponential expansion of quantum fluctuations in the very early ('Planck era') stages of the Universe, the scale-invariance being a result, in effect, of the self-similar nature of the exponential inflationary expansion. The other apparently inflation-supporting feature of the temperature variations lies in the fact that correlations are observed in these variations which seem to imply that separated regions of the sky must somehow have been in causal contact with one another, at some stage of the Universe's history, despite the fact that they would not be, in

[†] See, for example, Steinhardt and Turok (2007) where their cyclic cosmology is discussed in relation to this issue.

the standard (Friedmann–LeMaître–Robertson–Walker) cosmological models. The introduction of inflation brings these regions into causal contact.†

It should be mentioned that in spite of these successes, inflation makes no inroads into what I regard as the most fundamental mystery of the nature of the Big Bang, namely its *very* special geometry – resulting in our particular form of the Second Law – as seems to be well described by WCH. It is often argued that the very closely homogeneous and isotropic structure of the observed Universe is explained by inflation, but this is not so. Inflation by itself does not work as an explanation for this uniformity; its very operation (or that of inflation together with the 'anthropic principle'‡) requires the Universe to have been in an even *more* special configuration prior to the inflation, for the mechanism to operate at all.

It seems probable that this scale invariance and the apparently acausal correlations can be equally well accounted for within CCC. Both these features would be reflecting aspects of physical processes taking place in the aeon preceding our Big Bang, and I shall assume that this previous aeon is similar, in all relevant respects, to the aeon that we inhabit. Of course one must allow for a possibility that there *could* be major differences, perhaps resulting from those numbers that we call 'constants of Nature' taking values substantially different from those that we measure in our aeon. But in what follows, I shall assume that this is not the case. Accordingly, the nature of the previous aeon in *its* remote future should resemble what we expect of ours in *our* remote future.

It seems to me that the process taking place in the previous aeon of greatest relevance to the microwave temperature variations of our present aeon would be *gravitational wave bursts* arising from close encounters between black holes, as discussed earlier. In the vast periods of time before the eventual disappearance of black holes, due to

† See, for example, Guth (1981), Penrose (2004, §§28.3,4).

‡ See Penrose (1990), (2004, §§28.4,5,7).

Hawking evaporation, we expect very significant emissions of energy in the form of such radiation, containing a total amount of 'mass-energy'[†] at least comparable with that remaining in black holes and in unswallowed material. These bursts will leave their mark on $\mathcal{I}$, influencing its conformal 3-geometry and showing up in the *normal derivative* of the Weyl curvature across $\mathcal{I}$. This, according to CCC, will give rise to spatial variations in the density of material just after the Big Bang.

I shall come to the probable nature of this material, according to CCC, in Section 3.6, but irrespective of this, there appear to be certain characteristic features of the density variations whatever that material might be. In the first place, owing to the exponential nature of the expansion, it is a reasonable expectation that there should be a *scale-invariance* in these density variations. The argument is basically similar to that which is frequently used to derive such scale invariance from inflation, owing to an early-Universe exponential expansion that is argued for in that scheme of things. The difference is that in CCC, the exponential expansion took place *prior* to the Big Bang,[‡] not in the early stages following it.

Another distinction between CCC and inflation is that in the latter case the source of the density variations is taken to be 'quantum fluctuations' taking place in the early Universe, whereas in CCC the variations result from irregularities that are present in the late stages of the previous aeon. It seems to me that the main sources of such variations would arise from gravitational waves escaping to infinity (i.e. to the previous aeon's $\mathcal{I}$). These waves would give rise to a non-zero normal derivative in the Weyl curvature which, by CCC, would also be present at the $\mathcal{B}$ of the present aeon. This in turn, would lead (via the Bianchi identities referred to in Section 3.4) to non-zero tangential derivatives in the Ricci tensor **P**, i.e. by Einstein's equations, to density variations in the initial material.

[†] Remark due to C. W. Misner, see Penrose (1969).

[‡] See Veneziano (1998) for an earlier scheme involving this idea.

In CCC, such density variations ought to have an interesting structure. We take them to be built up from spherical patterns on the 3-surface $\mathcal{B}$, arising from the bursts of gravitational radiation that result from black-hole encounters in the previous aeon. These spheres would be places where the future light cones of such encounters intersect the $\mathcal{I}$ of the previous aeon. As viewed from our current spacetime location, we would see a superposition of circular patterns on the celestial sphere which could be likened (somewhat) to the appearance of ripples on a pond following a sustained period of rain. In principle, it should be possible to analyse the actual pattern of disturbance on the celestial sphere to see if it can be considered to be indeed a superposition of a large number of individual circles of disturbance. There would certainly be nonlocal correlations of the kind usually regarded as providing support for an early stage of inflationary expansion, but this detailed construction of the overall pattern in terms of circles would appear to be a particular signature of CCC. Nevertheless, some detailed and probably somewhat sophisticated analysis is needed, in order to see just what the explicit predictions of CCC should be, and to be reasonably certain whether or not the effect is present.

3.6 MORE FAR-REACHING AND SPECULATIVE ASPECTS OF CCC

But what kind of material would be involved in such density fluctuations? According to CCC, electromagnetic fields would propagate directly across the 3-surface connecting one aeon to the next, where the infinitely attenuated (and cold) electromagnetic radiation of the remote future of the earlier aeon would match precisely with the infinitely compressed (and hot) electromagnetic radiation of the next. In addition to this there would have to be a large proportion of new material most probably in the form of some fundamental scalar field ϕ (i.e. a spinless quantum particle), which is presumably not among the collection of possibilities laid out by the accepted standard model of particle physics. This ϕ-field would, in effect, be 'created' at the changeover from one aeon to the next, and this would come about from the conformal factor that gives the infinite changeover from the infinitely 'attenuated' metric of one aeon to the infinitely 'compressed' metric of the next.

It is part of the CCC scheme that the ϕ-field that appears to be necessarily introduced in the subsequent aeon could be interpreted as a creation of *dark matter* for that aeon. The dark matter would not be expected to be precisely conserved throughout any particular aeon, particularly because a good proportion of it would be swallowed by black holes, and these would finally be evaporated away. As distinct from the case of electromagnetism – where CCC would appear to demand a balance between the radiation in an aeon's remote future and that in the next big bang (so photon goes to photon in the transition from one aeon to the next, albeit with an infinite rescaling of the energy of each photon – the dark matter (or at least a good proportion of it) would, on the other hand, need to be regenerated at each transition. This appears to be not unreasonable, if we regard the 'flip-over' in the conformal factor at the transition to represent the creation of new dark matter (which, when one looks at the details, appears to be a mathematically consistent proposal).

This raises an issue that is a necessary feature of CCC also in other respects, namely the issue of the *massive* ingredients of the Universe that might be expected to persist right until the very remotest future, as represented by $\mathcal{I}$. So far, I have argued merely that a major proportion of the Universe's ingredients would be massless – and conformally invariant – entities, not that *all* such ingredients would be. Yet, it is really this stronger requirement that we need, if we are to be able to adopt the philosophical standpoint that the building of a clock is not possible in the very remote future, so that the conformal (causal) structure of spacetime becomes, in effect, the only relevant one. The mere presence of the occasional stray massive particle would seem to nullify this standpoint, so we need some way to address this issue.

When it comes to dark matter, we do not run into a conflict with observation or with standard particle physics if we are to assume that the dark-matter particles eventually decay into massless components. The decay lifetime can be extremely long, and there is certainly no conflict with observation here – or of theory, since it is completely unestablished what the dark matter actually is or how it fits in with the current Standard Model of particle physics. But what about the ordinary 'baryonic matter' that constitutes the material substance that we are familiar with? Well, it is certainly possible, within present-day proposals, that protons do eventually decay into less massive constituents, after decay times far longer than the present age of our Universe. What about the least massive particles known? Neutrinos are of three kinds and the *differences* between their three masses are known and are nonzero. But their absolute masses are not known, and there remains just a possibility that there is one neutrino type that is, in fact, massless, so that all other types eventually decay into it.

However, the most serious issue for CCC, in this respect, is the status of *electrons*. If we make the normal assumption that electric charge is exactly conserved in all physical interactions, then we would appear to conclude either that all electrons (being individually negatively charged) must eventually find positively charged partners, resulting in nothing but electrically neutral entities, or else that there are charged, ultimately massless, particles able to carry away the electric charge in a conformally invariant form. However, there cannot be massless charged particles in existence *now*, or else their potential presence would have become manifest in pair annihilation processes.[†] Moreover, a universe for which there is unending accelerated expansion (as CCC demands) would be expected to contain some 'rogue electrons' each of which would ultimately have no oppositely charged particles within its observable universe, so ultimate annihilation of its charge would not be possible. If such electrons remain massive,

[†] This point was stressed to me by James Bjorken.

then this would seem to violate the philosophical underpinnings of CCC, that only conformally invariant ingredients survive until the very remote future (in order that clocks cannot be constructed, in principle, near $\mathscr{I}$).

It seems to me that the least unpalatable possibility is that there is a gradual reduction of every electron's mass (and presumably of the mass of every other surviving particle), so that at $\mathscr{I}$ all particles, whether charged or not, indeed become eventually massless. This need not be inconsistent with observation, as the mass reduction could be extraordinarily slow, so that no hint of this would have been observed in experiments so far. However, there is a curious issue of principle that arises here, since, as we saw in Section 3.1, the very notion of time depends on the 'clocks' whose precise time rates come from the *masses* that the particles have. With an appropriate fall-off of the mass of electrons, the 'electron-age' of the entire Universe could be *finite*, the final bounding hypersurface $\mathscr{I}$ coming at a finite 'electron time'. There is the further possibility that other kinds of particles – say protons, assuming that they do not eventually decay into other particles (e.g. positrons and photons) – will also eventually lose mass to become ultimately massless, but perhaps at a different rate from electrons, so that the proton age of the Universe might come out to have a different finite value from the electron age. There does not seem to be anything inconsistent with this kind of possibility, and a deeper theory of particle physics than that which we have now would appear to be needed before such questions can be resolved.

There is one further possibility for providing an overall time scale, namely to use the *Planck mass*

$$m_{\text{Planck}} = \sqrt{\frac{c\hbar}{G}} \approx 2.1 \times 10^{-5}\,\text{g},$$

which is the absolute measure of mass associated with the Planck distance and Planck time introduced in Section 3.2. The actual building of a 'clock' based on this mass unit (according to the theoretical prescription of Section 3.1) would be enormously far from a practical matter, but the Planck mass has the theoretical advantage that it is not biased with respect to any particular choice of particle (e.g. electron, proton, etc.). In fact, the remoteness of the Planck mass from any actual 'clock' construction should be considered to be an advantage, in the present context, as it is being supposed that the relevant physical geometry in the very remote future is *conformal* spacetime geometry. Nevertheless it would seem to be not unreasonable to talk in terms of such Planck units when referring to the overall time scale of the Universe as being infinite.

There is another issue arising here which perhaps lends some credence to such an enormously slow decay of the mass values of actual physical particles, as measured against the Planck mass m_{Planck}. This is the fact that these actual values seem to 'make no sense' from the mathematical point of view. In fact, we find that at the present epoch, in terms of Planck units,

$$\text{mass of electron} \approx 4.3 \times 10^{-23} m_{\text{Planck}}$$
$$\text{mass of proton} \approx 7.9 \times 10^{-20} m_{\text{Planck}}.$$

There is no real understanding of such values within our present-day perspective on particle physics. If these numbers are not actually quite constant, albeit varying at such slow rates that the changes are completely undetected in current observations,[†] then we might regard the curiously tiny present values of today's particle masses, on the Planck scale, as mere accidents of our present 'date'!

Extrapolating in the opposite direction in time, we may be led to suspect that these particle-mass values could have been much greater, in terms of Planck mass values, at the vicinity of the Big Bang, and perhaps of the order of m_{Planck} itself. Yet, this possibility might be viewed as presenting a conflict with the underpinnings of the CCC scheme being proposed here, as it requires conformal invariance for all matter present at the Big Bang. The Universe (or present aeon) is often

[†] This scheme has some points in common with that of Dirac (1937, 1938) and also with that of Milne (1948), although both of these earlier schemes are in severe conflict with current observations. The version of CCC suggested in the text is very different from these in its observational implications.

considered to have 'started' at the time at which we might perhaps expect 'Planckian temperature' (of a Planck energy per degree of freedom, i.e. about 2.5×10^{32} K). But at that temperature it does not seem sensible to treat Planck-mass particles as massless, as would appear to be implied by the CCC requirements of conformal invariance at the Big Bang.

Yet, it seems to me that despite the common view that quantum-gravity considerations lead to an expectation that, at the Big Bang, Planck-scale physics ought to dominate, where a kind of chaotic 'quantum-foam' picture as referred to in Section 3.2 ought to hold, it is hard to see how this could be consistent with the strong kind of restriction that WCH implies, as seems to be demanded by the Second Law of thermodynamics. Indeed, CCC demands complete smoothness of the *conformal* geometry of spacetime at $\mathcal{B}$, as it joins on to the $\mathcal{I}$ of the previous aeon. There appears to be little role here for such a 'Planck-scale' physics dominating at the Big Bang, since it does not play such a role at the previous $\mathcal{I}$. Accordingly, it does not seem unreasonable to me that even if ordinary particles acquire Planck-scale masses at the Big Bang, the ultra-high energies near the Big Bang could still dominate, thereby still providing the effective conformal invariance at $\mathcal{B}$ that CCC requires.

It is curious that CCC seems to provide us with a scheme that is so much at variance with common quantum-gravity expectations with regard to spacetime singularities. But very little is really understood about Nature's actual role for 'quantum gravity' at the Big Bang in any case. Rather than a Planck-scale chaotic foam, the evidence seems to point, instead, to a highly ordered structure at the Big Bang, so the conformally smooth CCC picture might well provide a more compatible alternative.

On the other hand, the situation inside black holes has to be completely different from this. WCH provides no constraint whatever on the nature of the '*future*-type' spacetime singularities that arise within black holes. Here we must expect a Weyl curvature that *diverges to infinity*, and when radii of spacetime Weyl curvature begin to approach the Planck length of around 1.6×10^{-35} m, we must expect vast deviations from a smooth conformal geometry, as is indeed the implication of standard quantum-gravity arguments. Since we are so far from any actual quantum-gravity theory that is acceptable to all researchers in the field, it would not be unreasonable to form a pessimistic conclusion that "all bets are off" with regard to the quantum-gravity nature of black-hole singularities. However, the picture presented by Hawking evaporation, as discussed in Sections 3.3 and 3.4, suggests that all such Weyl-divergent 'future-type' singularities will have *disappeared* by the time $\mathcal{I}$ is reached, so the conformal smoothness demanded by CCC is not violated. There is, however, a key role for Hawking evaporation in CCC, and this has to do with the very issue that started us on this whole line of consideration, namely the Second Law of thermodynamics.

In the picture that CCC presents, we must have an exact matching between the available conformal physics at $\mathcal{I}$ (for the previous aeon) and that at $\mathcal{B}$ (for our own aeon). Let us take the relevant physics at $\mathcal{I}$ to be essentially just electromagnetic and gravitational radiation (as would be consistent with the picture presented in Section 3.4). The electromagnetic radiation is determined by the tensor **F**, and this simply carries through from the $\mathcal{I}$ of the previous aeon to the $\mathcal{B}$ of the current aeon. Using the normal derivative of **C** at the $\mathcal{I}/\mathcal{B}$ boundary to describe the gravitational radiation (which also contains the information of the intrinsic conformal 3-geometry of $\mathcal{I}/\mathcal{B}$), this also carries through from the previous aeon to the current one. In each case, the electromagnetic and the gravitational, there is an exact matching between the information on the previous $\mathcal{I}$ and the current $\mathcal{B}$. However, there is a difference in the physical interpretations in the gravitational case. At $\mathcal{I}$, this information is carried by **K**, which measures the ultimate gravitational radiation field, whereas at $\mathcal{B}$, the information is carried by the variation in the ϕ-field (presumably dark-matter density, in the viewpoint presented earlier) and the conformal 3-geometry.

The nature of this information (at either $\mathcal{I}$ or $\mathcal{B}$) determines the size and nature of the total phase-space available to a universe that acts in accordance with CCC. Since the information matches exactly from one side to the other, we may ask how this squares with the Second Law which, after all, was the starting point of this whole discussion. It would seem, at first sight, that nothing has been achieved, as there was as much freedom at the beginning of the Universe (our $\mathcal{B}$) as there will be at the end (our $\mathcal{I}$). The answer to this seeming conundrum is that there has to be huge loss of information (actually loss of phase-space volume) owing to degrees of freedom being absorbed in the future-type singularities of black holes. In Section 3.3, I addressed the issue of whether this information is *really* lost, rather than somehow being retrieved externally through hidden 'correlations' between internal and external states, as some physicists have argued for. The issue becomes particularly pertinent when the black hole eventually disappears – if in fact it actually does – through Hawking evaporation, and I expressed my own conviction earlier (Section 3.3) that this loss and disappearance must indeed be realities.

More than that, in the CCC picture, this loss of phase-space volume is *crucial* to the entire argument! It provides the reason that the initial state of each aeon indeed counts as very very special, and is represented as being restricted to a (relatively) extremely tiny volume $\mathcal{O}$ of allowable initial universe states, residing within the much larger phase space $\mathcal{U}$ which represents the possible alternatives that, without the WCH restriction, would have been available to the Universe (see Section 3.3). If we try to evolve the final states backwards in time – and let us denote the phase space of possible final states in the CCC picture by $\mathcal{F}$ – then we find that this backward-evolution is extremely far from unique, because of the information loss in black holes. The vast majority of these backward-evolved universes would be grossly inconsistent with WCH (and with the Second Law), leading to numerous white holes (the time reverses of black holes) which would have initial-type spacetime singularities with wildly diverging Weyl tensor. The idea is that the extremely 'special' Big Bang, and its attendant Second Law, as allowed by CCC, represents an extremely tiny selection of these backward evolutions.

But, the reader may well ask, how is it that an aeon's future evolution, consistent with the Second Law, can result in a universe state whose allowable phase space $\mathcal{F}$ is no larger than – and in fact *isomorphic* to – the allowable phase space $\mathcal{O}$ that the aeon was allowed to start from? The answer, according to CCC, lies in the non-intuitive fact that the phase space is continually being whittled down each time a black hole evaporates away, so that the operation of the Second Law is exactly balanced by this black-hole information loss.† One might be tempted to say that the black-hole evaporation represents a violation of the Second Law, but I believe that this is not the correct way of looking at it. The Hawking evaporation process remains consistent with the Second Law at all times (and indeed, such considerations formed part of the early discussions of this process‡); the point is that the phase space is effectively 'thinned out' all the time that a black hole absorbs information into its singularity (and the 'time' that this happens is itself ill defined, owing to the curious nature of the space-time geometry involved), during which process there is an effective reduction in the phase-space volume. The process is a subtle one, but it is not to be regarded as inconsistent with the Second Law.

Of course, the consistency of all these assertions must be checked in detail, in order that we may ascertain whether CCC provides a viable model for the Universe as we observe it. In particular, we may ask whether it is plausible that there is *enough* information loss in black holes to account for the enormous specialness of our actual Big Bang. In this regard, we must bear in mind that for a black hole that is basically in equilibrium with its surroundings, the information loss appears to be a continual process, and it cannot be estimated

† In earlier accounts, Penrose (1981, 2004, §30.9), I have tried to equate this 'information loss' – which is really phase-space reduction – to a phase-space gain due to an objective state reduction. I had never been able to make this argument quantitative, however, and it would seem that the issue is more complicated than I had previously envisaged, owing to the issues discussed in the text.

‡ See Hawking (1976); it should be pointed out that the evaporating away of a black hole, when the background temperature becomes lower than the hole's Hawking temperature, is indeed an entropy increasing process.

simply from its Bekenstein–Hawking entropy value. The longer that a black hole is in existence the greater will be the information loss, and this is relevant particularly to the extremely large and long-lived holes that settle in the centres of galactic clusters. All such consistency checks do, however, depend to some extent upon an assumption that each aeon is basically similar to the one preceding it. If the aeons are allowed to differ markedly from one another, then the scientific usefulness of the CCC proposal would be drastically reduced. But if the successive aeons are assumed to be basically similar to one another, then there are numerous constraints and consistency checks from both theory and observation, and this stronger version of CCC will stand or fall in accordance with them.

Acknowledgements

I am grateful to many of my colleagues for valuable discussions, notably to Florence Tsou, Hong-Mo Chan, Jacob Foster, James Bjorken, and especially to Paul Tod for valuable information concerning the relevant dynamical equations immediately following the Big Bang. I am grateful also to NSF for support under grant PHY00-90091.

BIBLIOGRAPHY

Arnol'd, V. I. (1978) *Mathematical Methods of Classical Mechanics*, Springer.

Ashtekar, A. (1986) New Hamiltonian formulation of general relativity, *Phys. Rev.* **D36**, 1587–1602.

Ashtekar, A. and Lewandowski, J. (2004) Background independent quantum gravity: a status report, *Class. Quantum Grav.* **21**, 53–152.

Bekenstein, J. (1993) How fast does information leak out of a black hole? *Phys. Rev. Lett.* **70**, 3680–3.

Braunstein, S. L. and Pati, A.K. (2007) Quantum information cannot be completely hidden in correlations: implications for the black-hole information paradox, *Phys. Rev. Lett.* **98**, 080502.

Connes, A. (1994) *Noncommutative Geometry*, Academic Press.

Diósi, L. (1984) Gravitation and quantum mechanical localization of macro-objects, *Phys. Lett.* **105A**, 199–202.

Diósi, L. (1989) Models for universal reduction of macroscopic quantum fluctuations, *Phys. Rev.* **A40**, 1165–74.

Dirac, P. A. M. (1937) The cosmological constants, *Nature*, **139**, 323.

Dirac, P. A. M. (1938) A new basis for cosmology, *Proc. Roy. Soc. (Lond.)* **A165**, 199.

Feynman, R. P., Morinigo, F. B. and Wagner, W. G. (1995) *The Feynman Lectures on Gravitation*, Addison-Wesley, §1.4, pp. 11–15.

Friedrich, H. (1998) Einstein's equation and conformal structure, in *The Geometric Universe: Science, Geometry, and the Work of Roger Penrose*, ed. S. A. Huggett, L. J. Mason, K. P. Tod, S. T. Tsou and N. M. J. Woodhouse, Oxford University Press.

Gibbons, G. W. and Hawking, S. W., eds. (1993) *Euclidean Quantum Gravity*, World Scientific.

Gibbons, G. W. and Perry, M. J. (1978) Black holes and thermal Green functions, *Proc. Roy. Soc. (Lond.)* **A358**, 467–94.

Guth, A. H. (1981) Inflationary universe: a possible solution to the horizon and flatness problems, *Phys. Rev.* **D23**, 347–56.

Hawking, S. W. (1975) Particle creation by black holes, *Comm. Math. Phys.* **43**, 199–220.

Hawking, S. W. (1976) Black holes and thermodynamics, *Phys. Rev.* **D13**(2), 191; Breakdown of predictability in gravitational collapse, *Phys. Rev.* **D14**, 2460.

Hawking, S. W. (2005) Information loss in black holes, *Phys. Rev.* **D72**, 084013-6.

Hawking, S. W. and Penrose, R. (1970) The singularities of gravitational collapse and cosmology, *Proc. Roy. Soc. (Lond.)* **A314**, 529–48.

Hodges, A. P. (1982) Twistor diagrams, *Physica* **114A**, 157–75.

Hodges, A. P. (1985) A twistor approach to the regularization of divergences, *Proc. Roy. Soc. (Lond.)* **A397**, 341–74; Mass eigenstatates in twistor theory, *ibid*, 375–96.

Hodges, A. P. (2006) Twistor diagrams for all tree amplitudes in gauge theory: a helicity-independent formalism [arXiv:hep-th/0512336v2]; Scattering amplitudes for eight gauge fields [arXiv:hep-th/0603101v1].

Kronheimer E. H. and Penrose, R. (1967) On the structure of causal spaces, *Proc. Camb. Phil. Soc.* **63**, 481–501.

Markopoulou, F. (1998) The internal description of a causal set: What the universe looks like from the inside, *Commun. Math. Phys.* **211**, 559-83. [gr-qc/9811053]

Marshall, W, Simon, C., Penrose, R. and Bouwmeester, D. (2003) Towards quantum superpositions of a mirror, *Phys. Rev. Lett.* **91**, 13–16; 130401. http://link.aps.org/abstract/prl/v91/e130401

Milne, E. A. (1948) *Kinematic Relativity*, Oxford University Press.

Penrose, R. (1964) Conformal approach to infinity, in *Relativity, Groups and Topology*: The 1963 Les Houches Lectures, ed. B. S. DeWitt and C. M. DeWitt, Gordon and Breach.

Penrose, R. (1965) Zero rest-mass fields including gravitation: asymptotic behaviour, *Proc. Roy. Soc. (Lond.)* **A284**, 159–203.

Penrose, R. (1967) Twistor algebra, *J. Math. Phys.* **8**, 345–66.

Penrose, R. (1969) Gravitational collapse: the role of general relativity, *Rivista del Nuovo Cimento* Serie I, **Vol. 1**; Numero speciale, 252–76. Reprinted: *Gen. Rel. Grav.* **34**, *no. 7, July 2002*, 1141–65.

Penrose, R. (1971) Angular momentum: an approach to combinatorial space-time, in *Quantum Theory and Beyond*, ed. Ted Bastin, Cambridge University Press.

Penrose, R. (1975) Twistor theory: its aims and achievements, in *Quantum Gravity, an Oxford Symposium*, eds. C. J. Isham, R. Penrose and D. W. Sciama, Oxford University Press.

Penrose, R. (1976) Non-linear gravitons and curved twistor theory, *Gen. Rel. Grav.* **7**, 31–52.

Penrose, R. (1979) Singularities and time-asymmetry, in *General Relativity: an Einstein Centenary*, ed. S. W. Hawking and W. Israel, Cambridge University Press.

Penrose, R. (1981) Time-asymmetry and quantum gravity, in *Quantum Gravity 2: A Second Oxford Symposium*, ed. D. W. Sciama, R. Penrose and C. J. Isham, Oxford University Press, pp. 244–72.

Penrose, R. (1986) Gravity and state-vector reduction, in *Quantum Concepts in Space and Time*, ed. R. Penrose and C. J. Isham, Oxford University Press, pp. 129–46.

Penrose, R. (1990) Difficulties with inflationary cosmology, in *Proceedings of the 14th Texas Symposium on Relativistic Astrophysics*, ed. E. Fenves, New York Academy of Sciences.

Penrose, R. (1996) On gravity's role in quantum state reduction, *Gen. Rel. Grav.* **28**, 581–600.

Penrose, R. (2000) Wavefunction collapse as a real gravitational effect, in *Mathematical Physics 2000*, ed. A. Fokas, T. W. B. Kibble, A. Grigouriou and B. Zegarlivoki, Imperial College Press, 266–82.

Penrose, R. (2004) *The Road to Reality: A Complete Guide to the Laws of the Universe*, Jonathan Cape.

Penrose, R. (2006) Before the big bang: an outrageous new perspective and its implications for particle physics, in *EPAC 2006 – Proceedings, Edinburgh, Scotland,* p. 2759 (organizer: C. Prior). (European Physical Society Accelerator Group, EPS-AG). http://accelconf.web.cern.ch/accelconf/e06/html/author.htm

Penrose, R. and MacCallum, M. A. H. (1972) Twistor theory: an approach to the quantization of fields and space-time, *Phys. Repts.* **6C**, 241–315.

Penrose, R. and Rindler, W. (1984) *Spinors and Space-Time, Vol. 1: Two-Spinor Calculus and Relativistic Fields*, Cambridge University Press.

Penrose, R. and Rindler, W. (1986) *Spinors and Space-Time, Vol. 2: Spinor and Twistor Methods in Space-Time Geometry*, Cambridge University Press.

Perlmutter, S. *et al.* (1998) Cosmology from type Ia supernovae, *Bull. Am. Astron.* **29**. [astro-ph/9812473]

Rugh, S. E. and Zinkernagel, H. (2007) Cosmology and the meaning of time (Symposium, 'The Socrates Spirit' Section for Philosophy and the Foundations of Physics, Hellabaekgade 27, Copenhagen N. Denmark).

Schild, A. (1949) Discrete space-time and integral Lorentz transformations, *Canad. J. Math.* **1**, 29–47.

Snyder, H. S. (1947) Quantized space-time, *Phys. Rev.* **71**, 38–41.

Sorkin, R. D. (1991) Spacetime and Causal sets, in *Relativity and Gravitation: Classical and Quantum*, ed. J. C. D'Olivo *et al.*, World Scientific.

Steinhardt. P. J. and Turok, N. (2007) *Endless Universe: Beyond the Big Bang*, Doubleday.

Taylor, J. H. (1998) Binary pulsars and relativistic gravity, in *The Universe Unfolding*, ed. H. Bondi and M. Weston-Smith, Oxford Clarendon Press.

Tod, K. P. (2003) Isotropic cosmological singularities: other matter models, *Class. Quantum Grav.* **20** 521–34 doi:10.1088/0264-9381/20/3/309.

Veneziano, G. (1998) A Simple/Short Introduction to Pre-Big-Bang Physics/Cosmology, in *Proc. of the 35th Course of the Erice School of Subnuclear Physics, "Highlights: 50 Years Later" (Erice, August 1997)*, ed. A. Zichichi.

Witten, E. (2004) Perturbative gauge theory as a string theory in twistor space, *Comm. Math. Phys.* **252**, 189.

4 On the fine structure of spacetime

Alain Connes

4.1 INTRODUCTION

Our knowledge of spacetime is described by two existing theories:

- General Relativity
- The Standard Model of particle physics

General Relativity describes spacetime as far as large scales are concerned and is based on the geometric paradigm discovered by Riemann. It replaces the flat (pseudo) metric of Poincaré, Einstein and Minkowski by a curved spacetime metric whose components form the gravitational potential. The basic equations are Einstein equations (Figure 4.1 in the absence of matter) which have a clear geometric meaning and are derived from a simple action principle. Many processes in physics can be understood in terms of an action principle, which says, roughly speaking, that the actual observed process minimises some functional, the action, over the space of possible processes. A simple example is Fermat's principle in optics which asserts that light of a given frequency traverses the path between two points which takes the least time. In the case of Einstein's equation the action is a functional on the set of all possible spacetime configurations and is computed by integrating the spacetime curvature over all of spacetime.

This Einstein–Hilbert action S_{EH}, from which the Einstein equations are derived in empty space, is replaced in the presence of matter by the combination

$$S = S_{EH} + S_{SM} \tag{4.1}$$

On Space and Time, ed. Shahn Majid. Published by Cambridge University Press.

FIG. 4.1 Einstein equations.

FIG. 4.2 CERN collision ring. Source: CERN.

where the second term S_{SM} is the Standard Model action which encapsulates our knowledge of all the different kinds of elementary particles to be found in nature. These are typically discovered by smashing together highly accelerated beams of other particles in huge particle accelerators such as at CERN (Figure 4.2) and what one finds, even at

FIG. 4.3 Standard Model Lagrangian.

the kinds of energies observable today (about 100 GeV†) is a veritable *zoo* of particles – electrons, quarks, neutrinos of various flavours to name a few – and their antiparticles as well as the force carriers (cf. Figure 4.4). The action S_{SM} encodes what is known so far about this array of particles and in principle describes all the matter and particles

† A GeV is a measure of energy equal to about 1.6×10^{-10} joules, or about the mass-energy of a proton.

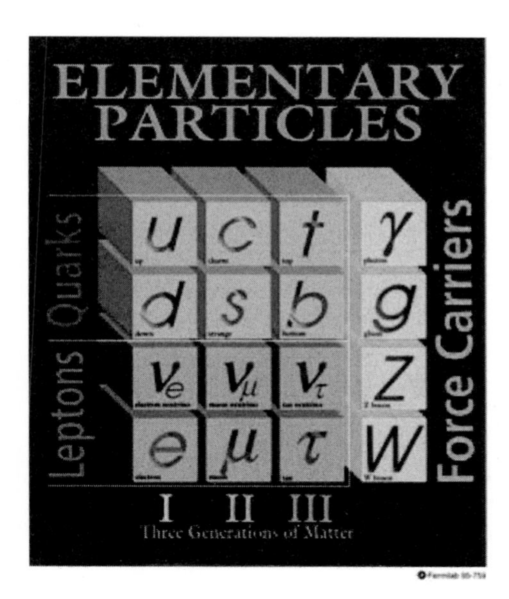

FIG. 4.4 Particles of the Standard Model. Source: Fermilab.

of force in the Universe, with the *exception* of gravity, to our best current understanding. It also has some twenty or more parameters put in by hand according to the experimental data, such as the masses of the various particles. The values of these parameters is a great mystery.

While the Einstein–Hilbert action S_{EH} has a clear geometric meaning, the additional term S_{SM} (cf. Figure 4.3) is quite complicated (it takes about four hours to typeset the formula) and is begging for a better understanding – in the words of an old friend:[†] it is a Shakespearean king disguised as a beggar.[‡]

Our goal in this essay is to explain that a conceptual understanding of the full action functional is now available[§] and shows that the additional term S_{SM} exhibits the fine texture of the geometry of spacetime. This fine texture appears as the product of the ordinary 4-dimensional continuum by a very specific finite discrete space F. Just to get a mental picture one may, in first approximation, think of

[†] Kastler (2000).
[‡] I told him 'I hope the beggar has diamonds in its pockets'.
[§] See Chamseddine, Connes and Marcolli (2006) and Chamseddine and Connes (2007).

F as a space consisting of two points. The product space then appears as a 4-dimensional continuum with 'two sides'. As we shall see, after a judicious choice of F, one obtains the full action functional (4.1) as describing *pure gravity* on the product space $M \times F$.

It is crucial, of course, to understand the 'raison d'être' of the space F and to explain why extending the ordinary continuum with such a space is necessary from first principles. As we shall see below such an explanation is now available.

4.2 THE QUANTUM FORMALISM AND ITS INFINITESIMAL VARIABLES

Let us start by explaining the formalism of quantum mechanics and how it provides a natural home for the notion of infinitesimal variable which is at the heart of the beginning of the 'calculus'.

One essential difference between the way Newton and Leibniz treated infinitesimals is that for Newton an infinitesimal is a *variable*. More precisely, according to Newton:†

> *In a certain problem, a variable is the quantity that takes an infinite number of values which are quite determined by this problem and are arranged in a definite order.*
>
> Each of these values is called a particular value of the variable.
>
> *A variable is called 'infinitesimal' if among its particular values one can be found such that this value itself and all following it are smaller in absolute value than an arbitrary given number.*

In the usual mathematical formulation of variables as maps from a set X to the real numbers $\mathbb{R}$, the set X has to be uncountable if some variable has continuous range. But then for any other variable with countable range some of the multiplicities are infinite. This means that discrete and continuous variables cannot coexist in this modern formalism.

† See A. N. Krylov, *Leonhard Euler*. Talk given on 5 October 1933, translated by N. G. Kuznetsov.

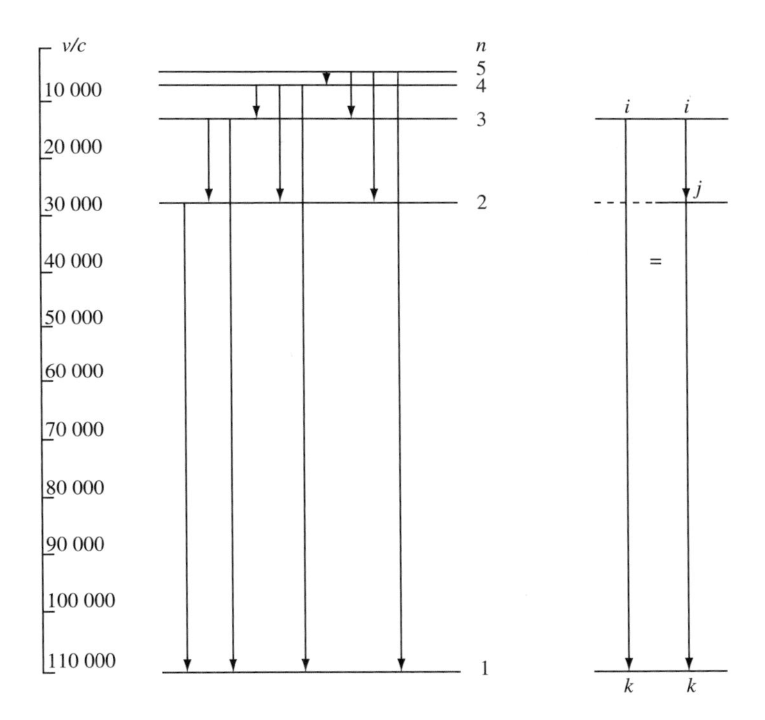

FIG. 4.5 Spectral lines and the Ritz–Rydberg law.

Fortunately everything is fine and this problem of treating continuous and discrete variables on the same footing is completely solved using the formalism of quantum mechanics which we now briefly recall.

At the beginning of the twentieth century a wealth of experimental data was being collected on the spectra of various chemical elements. These spectra obey experimentally discovered laws, the most notable being the Ritz–Rydberg combination principle (Figure 4.5). The principle can be stated as follows: spectral lines are indexed by pairs of labels. The statement of the principle then is that certain pairs of spectral lines, when expressed in terms of frequencies, do add up to give another line in the spectrum. Moreover, this happens precisely when the labels are of the form i, j and j, k, i.e. when the second label

(j) of the first spectral line coincides with the first label of the second spectral line.

In his seminal paper of 1925,[†] Heisenberg showed that this experimental law in fact dictates the algebraic rules for the observable quantities of the microscopic system given by an atom. It was Born who realised that what Heisenberg described in his paper corresponded to replacing classical coordinates with coordinates which no longer commute, but which obey the laws of matrix multiplication. In his own words reported in B. L. van der Waerden's book,[‡]

> After having sent Heisenberg's paper to the Zeitschrift für Physik for publication, I began to ponder about his symbolic multiplication, and was soon so involved in it that I thought the whole day and could hardly sleep at night. For I felt there was something fundamental behind it ... And one morning ... I suddenly saw light: Heisenberg's symbolic multiplication was nothing but the matrix calculus.

In other words, the Ritz–Rydberg law gives the groupoid law:

$$(i,\ j) \circ (j,\ k) = (i,\ k)$$

and the algebra of observables is given by the matrix product

$$(AB)_{ik} = \sum_j A_{ij} B_{jk},$$

for which in general commutativity is lost:

$$AB \neq BA. \tag{4.2}$$

This viewpoint on quantum mechanics was later somewhat obscured by the advent of the Schrödinger equation. The Schrödinger approach shifted the emphasis back to the more traditional technique

[†] Heisenberg (1925).
[‡] van der Waerden (1967).

of solving partial differential equations, while the more modern viewpoint of Heisenberg implied a much more serious change of paradigm, affecting our most basic understanding of the notion of space. We shall assume that the reader has some familiarity with the algebra of matrices of finite dimension. In quantum mechanics one needs an infinite-dimensional notion of these as linear operators on a 'Hilbert space'. The latter is formally defined as an infinite dimensional complex linear space with a positive-definite inner product.

Thus in quantum mechanics the basic change of paradigm has to do with the observable quantities which one would classically describe as real-valued functions on a set i.e. as a map from this set to real numbers. It replaces this classical notion of a 'real variable' by the following substitute: a self-adjoint operator in Hilbert space. Note that, if we were dealing with finite dimensional Hilbert spaces, a real variable would simply be a Hermitian matrix†, but to have more freedom, it is more convenient to take the Hilbert space to be the unique separable infinite-dimensional Hilbert space. All the usual attributes of real variables such as their range, the number of times a real number is reached as a value of the variable etc. have a perfect analogue in this quantum-mechanical setting. The range is the *spectrum* of the operator, and the spectral multiplicity gives the number of times a real number is reached.

In the early days of quantum mechanics, physicists had a clear intuition of this analogy between operators in Hilbert space (which they called q-numbers) and variables.

What is surprising is that the new set-up immediately provides a natural home for the 'infinitesimal variables'.

Indeed it is perfectly possible for an operator to be 'smaller than ϵ for any ϵ' without being zero. This happens when the size‡ of the restriction of the operator to subspaces of finite codimension tends to

† A matrix (A_{ij}) is Hermitian when $A_{ji} = \bar{A}_{ij}$ for all i, j.

‡ The size $\|T\|$ of an operator is the square root of the size of the spectrum of T^*T.

zero when these subspaces decrease (under the natural filtration by inclusion). The corresponding operators are called 'compact' and they share with naïve infinitesimals all the expected algebraic properties. In fact they comply exactly with Newton's definition of infinitesimal variables since the list of their characteristic values is a sequence decreasing to 0. All the expected algebraic rules such as infinitesimal $\times$ bounded $=$ infinitesimal etc. are fulfilled. The only property of the naïve infinitesimal calculus that needs to be dropped is the commutativity.

It is only because one drops commutativity that variables with continuous range can coexist with variables with countable range. In the classical formulation of variables, as maps from a set X to the real numbers, we saw above that discrete variables cannot coexist with continuous variables. The uniqueness of the separable infinite dimensional Hilbert space cures that problem,† and variables with continuous range coexist happily with variables with countable range, such as the infinitesimal ones. The only new fact is that they do not commute.

One way to understand the transition from the commutative to the noncommutative is that in the latter case one needs to care about the ordering of the letters when one is writing,‡ whereas the commutative rule oversimplifies the computations. As explained above, it is Heisenberg who discovered that such care was needed when dealing with the coordinates on the phase space of microscopic systems.

At the philosophical level there is something quite satisfactory in the variability of quantum-mechanical observables. Usually when pressed to explain what is the cause of the variability in the external world, the answer that comes naturally to the mind is just: *the passing of time*. But precisely the quantum world provides a more subtle answer since the *reduction of the wave packet* which happens in any

† Since for instance the Hilbert space $L^2([0,1])$ is the same as $\ell^2(\mathbb{N})$.

‡ As an example use the 'commutative rule' to simplify the following cryptic message I received from a friend: 'Je suis alenconnais, et non alsacien. Si t'as besoin d'un conseil nana, je t'attends au coin annales. Qui suis-je?'

quantum measurement is nothing else but the replacement of a 'q-number' by an actual number which is chosen among the elements in its spectrum. Thus there is an intrinsic 'variability' in the quantum world which is so far not reducible to anything classical. The results of observations are intrinsically *variable* quantities, and this to the point that their values cannot be reproduced from one experiment to the next, but which, when taken altogether, form a q-number.

Heisenberg's discovery shows that the phase-space of microscopic systems is noncommutative inasmuch as the coordinates on that space no longer satisfy the commutative rule of ordinary algebra. This example of the phase space can be regarded as the historic origin of noncommutative geometry. But what about spacetime itself ? We now show why it is a natural step to extend usual commutative spacetime to a noncommutative one.

4.3 WHY SPACETIME SHOULD BE EXTENDED TO A NONCOMMUTATIVE ONE

The full action (4.1) of gravity coupled with matter

$$S = S_{\mathrm{EH}} + S_{\mathrm{SM}}$$

admits a huge natural group of symmetries. The group of invariance for the Einstein–Hilbert action S_{EH} is the group of diffeomorphisms of the manifold M and the invariance of the action S_{EH} is simply the manifestation of its geometric nature. A diffeomorphism acts by permutations of the points of M so that points have no absolute meaning. The full group $\mathcal{U}$ of invariance of the action (4.1) is, however, richer than the group of diffeomorphisms of the manifold M since one needs to include something called 'the group of gauge transformations' which physicists have identified as the symmetry of the matter part S_{SM}. This is defined as the group of maps from the manifold M to some fixed other group, G, called the 'gauge group', which as far as we know is:

$$G = U(1) \times SU(2) \times SU(3).$$

The group of diffeomorphisms acts on the group of gauge transformations by permutations of the points of M and the full group of symmetries of the action S is the semi-direct product of the two groups (in the same way, the Poincaré group, which is the invariance group of Special Relativity, is the semi-direct product of the group of translations by the group of Lorentz transformations). In particular it is not a simple group† but is a 'composite' and contains a huge normal subgroup.

Now that we know the invariance group $\mathcal{U}$ of the action (4.1), it is natural to try and find a space X whose group of diffeomorphisms is simply that group, so that we could hope to interpret the full action as pure gravity on X. This is an old 'Kaluza–Klein' idea in physics. Unfortunately this search is bound to fail if one looks for an ordinary manifold since by a mathematical result the connected component of the identity in the group of diffeomorphisms is always a simple group, excluding a semi-direct product structure such as that of $\mathcal{U}$.

But noncommutative spaces of the simplest kind readily give the answer, modulo a few subtle points. To understand what happens note that for ordinary manifolds the algebraic object corresponding to a diffeomorphism is just an automorphism of the algebra of coordinates i.e. a transformation of the coordinates that does not destroy their algebraic relations. When an algebra is not commutative there is an easy way to construct automorphisms. One takes an element u of the algebra and one assumes that u is invertible i.e. its inverse u^{-1} exists, solution of the equation $uu^{-1} = u^{-1}u = 1$. Using u one obtains a so-called *inner* automorphism, by the formula

$$\alpha(x) = u x u^{-1}, \quad \forall x \in \mathcal{A}.$$

Note that in the commutative case this formula just gives the identity automorphism (since one could then permute x and u^{-1}). Thus this construction is interesting only in the noncommutative case.

† A simple group is one which cannot be decomposed into smaller pieces, a bit like a prime number cannot be factorised into a product of smaller numbers.

Moreover the inner automorphisms form a subgroup $\text{Int}(\mathcal{A})$ which is always a normal subgroup of the group of automorphisms.†

In the simplest example, where we take for $\mathcal{A}$ the algebra of smooth maps from a manifold M to the algebra of $n \times n$ matrices of complex numbers, one shows that the group $\text{Int}(\mathcal{A})$ in that case is (locally) isomorphic to the group of gauge transformations i.e. of smooth maps from M to the gauge group $G = PSU(n)$ (quotient of $SU(n)$ by its centre). Moreover the relation between inner automorphisms and all automorphisms becomes identical to the exact sequence governing the structure of the group $\mathcal{U}$.

It is quite striking that the terminology of internal symmetries coming from physics agrees so well with the mathematical one of inner automorphisms. In the general case only automorphisms that are unitarily implemented in Hilbert space will be relevant but modulo this subtlety one can see at once from the above example the advantage of treating noncommutative spaces on the same footing as the ordinary ones. The next step is to properly define the notion of metric for such spaces and we shall first indulge in a short historical description of the evolution of the definition of the 'unit of length' in physics. This will prepare the ground for the introduction to the spectral paradigm of noncommutative geometry in the following section.

4.4 A BRIEF HISTORY OF THE METRIC SYSTEM

The next step is to understand what is the replacement of the Riemannian paradigm for noncommutative spaces. To prepare for that we now tell the story of the change of paradigm that already took place in the metric system with the replacement of the concrete 'mètre étalon' by a spectral unit of measurement.

We describe the corresponding mathematical paradigm of noncommutative geometry in Section 4.5.

† Since we deal with involutive algebras we use unitaries: $u^{-1} = u^*$ to define inner automorphisms.

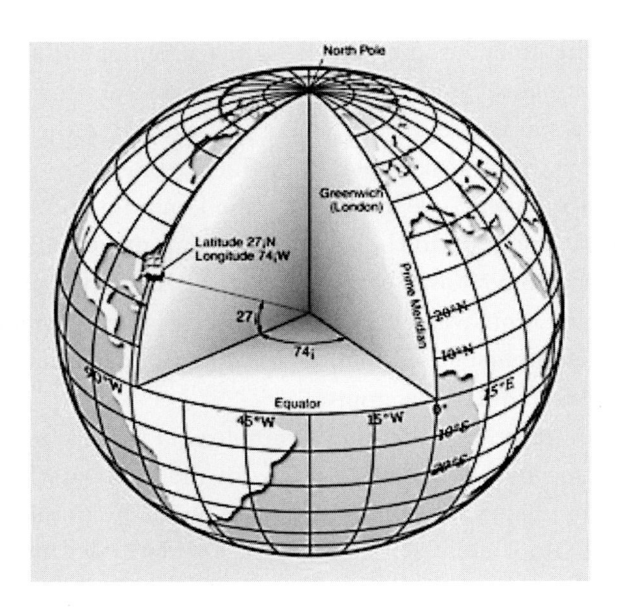

FIG. 4.6 Quarter of meridian.

The notion of geometry is intimately tied up with the measurement of length. In the real world such measurement depends on the chosen system of units and the story of the most commonly used system – the metric system – illustrates the difficulties attached to reaching some agreement on a physical unit of length which would unify the previous numerous existing choices.

In 1791 the French Academy of Sciences agreed on the definition of the unit of length in the metric system, the 'mètre', as being 10^{-7} times the quarter of the meridian of the Earth (Figure 4.6). The idea was to measure the length of the arc of the meridian from Barcelona to Dunkirk while the corresponding angle (approximately 9.5) was determined using the measurement of latitude from reference stars. In a way this was just a refinement of what Eratosthenes had done in Egypt, 250 years BC, to measure the size of the Earth (with a precision of 0.4 per cent). Thus in 1792 two expeditions were sent to measure this arc of the meridian; one for the northern portion was led by

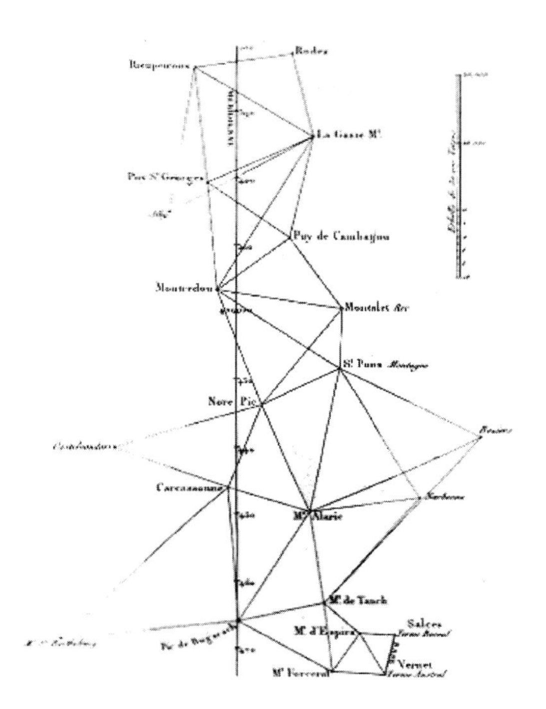

FIG. 4.7 The method of triangulation.

Delambre and the other for the southern portion was led by Méchain. Both of them were astronomers who were using a new instrument for measuring angles, invented by Borda, a French physicist. The method they used is the method of triangulation (Figure 4.7) and of concrete measurement of the 'base' of one triangle. It took them a long time to perform their measurements and it was a risky enterprise. At the beginning of the revolution, France entered into a war with Spain. Just try to imagine how difficult it is to explain that you are trying to define a universal unit of length when you are arrested at the top of a mountain with very precise optical instruments allowing you to follow all the movements of the troops in the surroundings. Both Delambre and Méchain were trying to reach the utmost precision in their measurements and an important part of the delay came from the fact that this reached an obsessive level in the case of Méchain. In fact when he measured the latitude of Barcelona he did it from two different close-by locations, but found contradictory results which were discordant by 3.5 seconds of arc. Pressed to give his result he chose to hide this discrepancy just to 'save face', which is the wrong attitude for a scientist. Chased from Spain by the war with France he had no second chance to understand the origin of the discrepancy and had to fiddle a little with his results to present them to the International Commission which met in Paris in 1799 to collect the results of Delambre and Méchain and compute the 'mètre' from them. Since he was an honest man obsessed by precision, the above discrepancy kept haunting him and he obtained permission from the Academy to lead another expedition a few years later to triangulate further into Spain. He went, and died from malaria in Valencia. After his death, his notebooks were analysed by Delambre who found the discrepancy in the measurements of the latitude of Barcelona but could not explain it. The explanation was found 25 years after the death of Méchain by a young astronomer by the name of Nicollet, who was a student of Laplace. Méchain had done in both of the sites he had chosen in Barcelona (Mont Jouy and Fontana del Oro) a number of measurements of latitude using several reference stars. Then he had simply taken the average of his measurements in each place. Méchain knew very well that refraction distorts the path of light rays which creates an uncertainty when you use reference stars that are close to the horizon. But he considered that the average result would wipe out this problem. What Nicollet did was to ponder the average to eliminate the uncertainty created by refraction and, using the measurements of Méchain, he obtained a remarkable agreement (0.4 seconds i.e. a few metres) between the latitudes measured from Mont Jouy and Fontana del Oro. In other words Méchain had made no mistake in his measurements and could have understood by pure thought what was wrong in his computation. I recommend the book of Ken Adler for a nice account of the full story of the two expeditions.† In any case in the

† Adler (2002).

FIG. 4.8 Modern atomic clocks under development are the size of chips. Credit: J. Kitching/NIST. The first accurate atomic clock was built in 1955 and led to the internationally agreed definition of the second being based on atomic time. The second is currently defined as the duration of 9 192 631 770 periods of the radiation corresponding to the transition between the two hyperfine levels of the caesium atom.

meantime the International Commission had taken the results from the two expeditions and computed the length of the ten millionth part of the quarter of the meridian using them. Moreover, a concrete platinum bar with approximately that length was then realised and was taken as the definition of the unit of length in the metric system. With this unit the actual length of the quarter of meridian turns out to be 10 002 290 rather than the aimed for 10 000 000 but this is no longer relevant. In fact in 1889 the reference became another specific metal bar (of platinum and iridium) which was deposited near Paris in the pavillon de Breteuil. This definition held until 1960.

Already in 1927, at the seventh conference on the metric system, in order to take into account the inevitable natural variations of the concrete 'mètre-étalon', the idea emerged to compare it with a reference wavelength (the red line of cadmium). Around 1960 the reference to the 'mètre-étalon' was finally abandoned and a new definition of the 'mètre' was adopted as 1 650 763.73 times the wavelength of the radiation corresponding to the transition between the levels 2p10 and 5d5 of krypton 86Kr. In 1967 the second was defined as the duration of 9 192 631 770 periods of the radiation corresponding to the transition

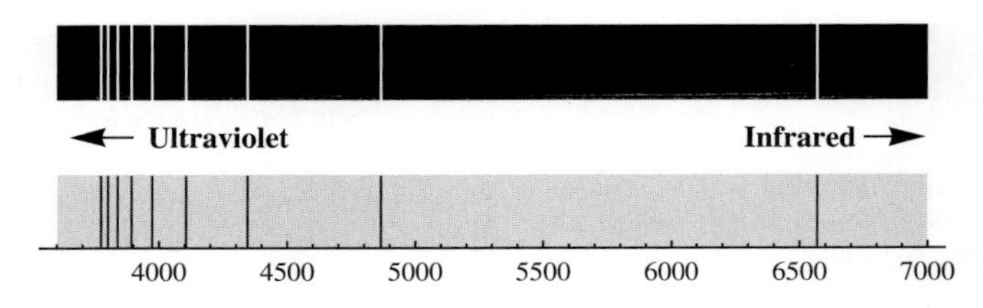

FIG. 4.9 Wavelengths of spectral lines for hydrogen.

between the two hyperfine levels of caesium-133. Finally in 1983 the 'mètre' was defined as the distance travelled by light in 1/299 792 458 seconds. In fact the speed of light is just a conversion factor and to define the 'mètre' one gives it the specific value of

$$c = 299\,792\,458\,\text{m/s}.$$

In other words the 'mètre' is defined as a certain fraction $\frac{9\,192\,631\,770}{299\,792\,458} \sim 30.6633\ldots$ of the wavelength of the radiation coming from the transition between the above hyperfine levels of the caesium atom.

The advantages of the new standard of length are many. First by not being tied up with any specific location it is in fact available anywhere where caesium is. The choice of caesium as opposed to helium or hydrogen which are much more common in the Universe is of course still debatable, and it is quite possible that a new standard will soon be adopted involving spectral lines of hydrogen instead of caesium.

While it would be difficult to communicate our standard of length with other extraterrestrial civilisations if they had to make measurements of the Earth (such as its size) the spectral definition can easily be encoded in a probe and sent out. In fact spectral patterns (Figure 4.9) provide a perfect 'signature' of chemicals, and a universal information available anywhere where these chemicals can be found, so that the wavelength of a specific line is a perfectly acceptable unit.

4.5 SPECTRAL GEOMETRY

It is natural to wonder whether one can adapt the basic paradigm of geometry to the new standard of length, described in the previous section, which is of spectral nature. The Riemannian paradigm is based on the Taylor expansion in local coordinates x^μ of the square of the line element†, in the form

$$ds^2 = g_{\mu\,\nu}\,\mathrm{d}x^\mu\,\mathrm{d}x^\nu \tag{4.3}$$

where the notation implies a sum over the labels μ, ν from 1 up to the dimension of the manifold. The measurement of the distance between two points is given by the geodesic formula

$$d(A,B) = \mathrm{Inf}\int_\gamma \sqrt{g_{\mu\,\nu}\,\mathrm{d}x^\mu\,\mathrm{d}x^\nu} \tag{4.4}$$

where the infimum‡ is taken over all paths γ from A to B. A striking feature of this formula (4.4) for measuring distances is that it involves taking a square root. It is often true that 'taking a square root' in a brutal manner as in (4.4) is hiding a deeper level of understanding. In fact this issue of taking the square root led Dirac to his famous analogue of the Schrödinger equation for the electron and the theoretical discovery of the positron. Dirac was looking for a relativistic invariant form of the Schrödinger equation. One basic property of that equation is that it is of first order of differentiation in the time variable. The Klein–Gordon equation, which is the relativistic form of the Laplace equation, is a relativistic invariant but is of second order of differentiation in time. Dirac found a way to take the square root of the Klein–Gordon operator using Clifford algebra. In fact (as pointed out to me by Atiyah) Hamilton had already written the required magical expression using his quaternions. Meanwhile, when I was in St Petersburg for Euler's 300th, I noticed that Euler could almost be credited for quaternions since he had explicitly written their multiplication

† Note that one should not confuse the 'line element' ds with the unit of length. In the classical framework, the latter allows one to give a numerical value to the distance between nearby points in the form (4.3). Multiplying the unit of length by a scalar λ one divides the line element ds by λ since ds is measured by its ratio with the unit of length.

‡ The infimum of a collection of numbers differs from the minimum in that it need not itself be a member of the collection. The supremum similarly differs from maximum only by this subtlety.

rule in order to show that the product of two sums of 4 squares is a sum of 4 squares.

So what is the relation between Dirac's square root of the Laplacian and the above issue of taking the square root in the formula for the distance $d(A, B)$? The point is that one can use Dirac's solution and rewrite the same distance function (4.4) in the following manner,

$$d(A, B) = \text{Sup}\{|f(A) - f(B)|\,;\ f \in \mathcal{A},\ \|[D, f]\| \leq 1\} \tag{4.5}$$

where one uses the quantum set-up of Section 4.2. Here $[D, f]$ denotes the commutator $[D, f] = Df - fD$ of the two operators and $\|[D, f]\|$ is its size (as explained in a footnote above). The space X is encoded through the algebra $\mathcal{A}$ of functions on X and, as in Section 4.2, this algebra is concretely represented as operators in Hilbert space $\mathcal{H}$, which is here the Hilbert space of square integrable spinors. The ingredient that allows us to measure distances is the operator D, which is the Dirac operator in the above case of ordinary Riemannian geometry. The new formula (4.5) gives the same result as the geodesic distance formula (4.4) but it is of a quite different nature. Indeed, instead of drawing a path from A to B in our space, i.e. a map from the interval $[0, 1]$ to X, we use maps from X to the real line. In general for a given space X it is not possible to join any two points by a continuous path, and thus the formula (4.4) is limited in its applications to spaces which are *arcwise connected*. On the opposite side, by a theorem of topology any compact space X admits plenty of continuous functions and the corresponding algebra of continuous functions can be concretely represented in Hilbert space. Thus the new formula (4.5) can serve as the basis for the measurement of distances and hence of the geometry of the space, provided we fix the operator D. Now what is the intuitive meaning of D? Note that formula (4.5) is based on the lack of commutativity between D and the coordinates f on our space. Thus there should be a tension that prevents D from commuting with the coordinates. This tension is provided by the following key hypothesis '*the inverse of D is an infinitesimal*'. Indeed we saw in Section 4.2 that variables with continuous range cannot commute with infinitesimals,

FIG. 4.10 Paul Dirac.

which gives the needed tension. But there is more, because of the following equation which gives to the inverse of D the heuristic meaning of the *line element*:†

$$ds = D^{-1}.$$

Thus one can think of a geometry as a concrete Hilbert space representation not only of the algebra of coordinates on the space X we are interested in, but also of its infinitesimal line element ds. In the usual Riemannian case this representation is moreover irreducible. Thus in many ways this is analogous to thinking of a particle as Wigner taught us, i.e. as an irreducible representation (of the Poincaré group).

One immediate advantage of this point of view on geometry is that it uses directly the algebra of coordinates on the space X one is interested in, rather than its set-theoretic nature. In particular it no longer demands that this algebra be commutative.

There are very natural geometric spaces such as, for instance, the space of geodesics of a Riemann surface, or more generally spaces of leaves of foliations which even though they are 'sets' are better

† Note that D has the dimension of the inverse of a length.

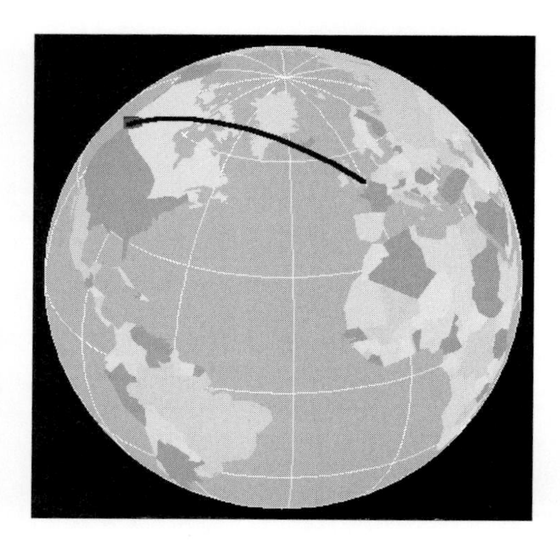

FIG. 4.11 Geodesic path.

described by a noncommutative algebra. The point is simple to understand: these spaces are constructed by the process of passing to the quotient and this cannot be done at once since it is impossible to select a representative in each equivalence class. Thus the quotient is encoded by the equivalence relation itself and this generates the noncommutativity, exactly as in the case of the original discovery of Heisenberg from the Ritz–Rydberg composition law. In case it is possible to select a representative in each equivalence class (as it is for instance for the geodesics on the sphere of Figure 4.11), the obtained algebra is equivalent (in a sense called 'Morita equivalence' on which we shall elaborate a bit below) to a commutative one, so that nothing is lost in simplicity. But one can now comprehend algebraically spaces of geometric nature which would seem intractable otherwise, a simple example being the space of geodesics on a torus, or even simpler the subspace obtained by restricting to those geodesics with a given slope. One obtains in this way the noncommutative torus, hard to 'visualise' but whose subtleties are perfectly accessible through algebra and analysis.

In noncommutative geometry a space X is described by the corresponding algebra $\mathcal{A}$ of coordinates which is now no longer assumed to be commutative i.e. by an involutive algebra $\mathcal{A}$ concretely represented as operators in Hilbert space $\mathcal{H}$ and the line element ds which is an infinitesimal, concretely represented in the same Hilbert space $\mathcal{H}$. Equivalently a noncommutative geometry is given by a spectral triple $(\mathcal{A}, \mathcal{H}, D)$, where D is the inverse of ds. Thus in noncommutative geometry the basic classical formula (4.4) is replaced by (4.5) where D is the inverse of the line element ds.

The new paradigm of spectral triples passes a number of tests to qualify as a replacement of Riemannian geometry in the noncommutative world:

- It contains the Riemannian paradigm as a special case.
- It does not require the commutativity of coordinates.
- It covers the spaces of leaves of foliations.
- It covers spaces of fractal, complex or infinite dimension.
- It applies to the analogue of symmetry groups (compact quantum groups).
- It provides a way of expressing the full Standard Model coupled to Einstein gravity as pure gravity on a modified spacetime geometry.
- It allows for quantum corrections to the geometry.

The reason why the quantum corrections can be taken into account is that the physics meaning of the line element ds is as the propagator for fermions, which receives quantum corrections in the form of 'dressing'. The fact that the paradigm of spectral triples is compatible with compact quantum groups is the result of a long saga.

The traditional notions of geometry all have natural analogues in the spectral framework. We refer to Connes and Marcolli (2007) for more details. Some notions become more elaborate, such as that of dimension. The dimension of a noncommutative geometry is not a number but a spectrum, the *dimension spectrum*, which is the subset of the complex plane at which the spectral functions have singularities.

Noncommutativity also brings new features which have no counterpart in the commutative world. In fact we already saw this in the case of the inner automorphisms which are trivial in the commutative case but correspond to the internal symmetries in physics in the noncommutative case. A similar phenomenon occurs also at the level of the metric i.e. of the operator D. To understand what happens one needs to realise first that some noncommutative algebras can be intimately related to each other without being the same. For instance an algebra $\mathcal{A}$ shares most of its properties with the algebra $M_n(\mathcal{A})$ of $n \times n$ matrices over $\mathcal{A}$. Of course if $\mathcal{A}$ is commutative this is no longer the case for $M_n(\mathcal{A})$, but the theory of 'vector spaces† over $\mathcal{A}$' is unchanged when one replaces $\mathcal{A}$ by $M_n(\mathcal{A})$ or more generally by the algebra $\mathcal{B}$ of endomorphisms of such a vector space over $\mathcal{A}$. This relation between noncommutative algebras is called Morita equivalence and gives an equivalence relation between algebras. At the intuitive level two Morita equivalent algebras describe the same noncommutative space. Given a spectral geometry $(\mathcal{A}, \mathcal{H}, D)$ one would like to transport the geometry to a Morita equivalent algebra $\mathcal{B}$ as above. This is easily done for the Hilbert space, but when one tries to transport the metric, i.e. the operator D, one finds that there is a choice involved: that of a 'connection' A. Any algebra is Morita equivalent to itself and the above ambiguity generates a bunch of metrics D_A which are 'internally related' to the given one D. In this way one gets the inner deformations of the spectral geometry, which will account for the gauge bosons in the physics context.

So much for the 'metric' aspect but what about the choice of sign which is often involved when one is taking a square root and hence should play a role here since our line element ds is based on Dirac's square root for the Laplacian. Now this choice of sign, which amounts in the case of ordinary geometry to the choice of a spin structure which is the substitute for the choice of an orientation, turns out to be deeply related to the very notion of 'manifold'. While one can have a good recipe for constructing manifolds by gluing charts together, it is far less obvious to understand in a conceptual manner the global properties which characterise the spaces underlying manifolds. For instance the same homotopy type can underlie quite different manifolds which

† In technical terms one deals with finite projective modules over $\mathcal{A}$.

are not homeomorphic to each other and are distinguished by the fundamental invariant given by the Pontrjagin classes. In first approximation, neglecting subtleties coming from the role of the fundamental group and also of the special properties of dimension 4, the choice of a manifold in a given homotopy type is only possible if a strong form of Poincaré duality holds.† Usual Poincaré duality uses the fundamental class in ordinary homology to yield an isomorphism between homology and cohomology.‡ The more refined form of Poincaré duality that is involved in the choice of a manifold in a given homotopy type involves a finer homology theory called KO-homology. The operator theoretic realisation of cycles in KO-homology was pioneered by Atiyah and Singer as a byproduct of their Index Theorem and has reached a definitive form in the work of Kasparov. The theory is periodic with period 8 and one may wonder how a number modulo 8 can appear in its formulation. This manifests itself in the form of a table of signs which governs the commutation relations between two simple

n	0	1	2	3	4	5	6	7
ε	1	1	-1	-1	-1	-1	1	1
ε'	1	-1	1	1	1	-1	1	1
ε''	1		-1		1		-1	

decorations of spectral triples (with the second only existing in the even dimensional case) so that they yield a KO-homology class:

(i) An antilinear isometry J of $\mathcal{H}$ with $J^2 = \epsilon$ and $JD = \epsilon' DJ$.
(ii) A $\mathbb{Z}/2$-grading γ of $\mathcal{H}$, such that $J\gamma = \epsilon'' \gamma J$.

The three signs $(\epsilon, \epsilon', \epsilon'') \in \{\pm 1\}^3$ detect the dimension modulo 8 of the cycle in KO-homology, as ruled by the above table.

† Cf. Sullivan (2005).

‡ The homology and cohomology groups associated to a manifold provide information about its structure. The former can be constructed internally to the manifold by means of 'cycles', while the latter can be constructed, for example, in terms of differentials on the manifold.

These decorations have a deep meaning both from the mathematical point of view – where the main underlying idea is that of a manifold – as well as from the physics point of view. In the physics terminology the operator J is the *charge conjugation* operator, and the grading γ is the *chirality*.

The compatibilities of these simple decorations with the algebra $\mathcal{A}$ are the following: the antilinear involution J fulfils the 'order zero' condition

$$[a, J b^* J^{-1}] = 0, \quad \forall a, b \in \mathcal{A}, \tag{4.6}$$

and the $\mathbb{Z}/2$-grading γ fulfils $\gamma \mathcal{A} \gamma^{-1} = \mathcal{A}$, which gives a $\mathbb{Z}/2$-grading of the algebra $\mathcal{A}$. Here [,] denotes 'commutator' defined as product minus product in the reverse order.

The mathematical origin of the commutation relation (4.6) is in the work of the Japanese mathematician M. Tomita who proved that, given an involutive algebra $\mathcal{A}$ of operators in $\mathcal{H}$, one obtains an antilinear involution J fulfilling (4.6) by taking the polar decomposition of the operator

$$a\xi \mapsto a^*\xi, \quad \forall a \in \mathcal{A}$$

under the key hypothesis that the vector ξ is *separating* for $\mathcal{A}$ i.e. that $\mathcal{A}'\xi = \mathcal{H}$ where the commutant $\mathcal{A}'$ is the algebra of operators that commute with every element of $\mathcal{A}$.

Thus, to summarise, a *real* spectral triple (i.e. a spectral triple $(\mathcal{A}, \mathcal{H}, D)$ together with the decorations J and γ) encodes both the metric and the fundamental class in KO-homology for a noncommutative space. A key condition that plays a role in characterising the classical case is the following 'order one' condition relating the operator D with the algebra:

$$[[D, a], b^0] = 0 \qquad \forall a, b \in \mathcal{A} \tag{4.7}$$

where we use the notation $b^0 = J b^* J^{-1}$ for any $b \in \mathcal{A}$.

4.6 THE 'RAISON D'ÊTRE' OF THE FINITE SPACE F

Poincaré and Einstein showed how to infer the correct flat spacetime geometry (known as Minkowski space geometry) underlying Special Relativity from experimental evidence associated to Maxwell's theory of electromagnetism. In fact, the Maxwell equations are intrinsically relativistic. This flat geometry is then extended to the curved Lorentzian manifolds of General Relativity. From the particle physics viewpoint, the Lagrangian of electromagnetism is however just a very small part of the full Standard Model Lagrangian (cf. Figure 4.3). Thus, it is natural to wonder whether the transition from the Lagrangian of electrodynamics to the Standard Model can be understood as a further refinement of the geometry of spacetime, rather than the introduction of a zoo of new particles.

The spectral formalism of noncommutative geometry, explained above, makes it possible to consider spaces which are more general than ordinary manifolds. This gives us more freedom to obtain a suitable geometric setting that accounts for the additional terms in the Lagrangian.

The idea to obtain the spectral geometry $(\mathcal{A}, \mathcal{H}, D)$ is that:

- The algebra $\mathcal{A}$ is dictated by the comparison of its group of inner automorphisms with the internal symmetries.
- The Hilbert space $\mathcal{H}$ is the Hilbert space of Euclidean fermions.
- The line element ds is the propagator for Euclidean fermions.

Using this idea together with the spectral action principle,[†] which will be explained in the next section, allowed to determine a very specific finite noncommutative geometry F such that pure gravity (in the form of the spectral action) on the product $M \times F$, with the metric given by the inner fluctuations of the product metric, delivers the Standard Model coupled to gravity.

Thus in essence what happens is that the scrutiny of spacetime at very small scales (of the order of 10^{-16} cm) reveals a *fine structure*

[†] Chamseddine and Connes (1996).

which replaces an ordinary point in the continuum by the finite geometry F.

In the above approach this finite geometry was taken from the phenomenology i.e. put in by hand to obtain the Standard Model Lagrangian using the spectral action. The algebra $\mathcal{A}_F$, the Hilbert space $\mathcal{H}_F$ and the operator D_F for the finite geometry F were all taken from the experimental data. The algebra comes from the gauge group, the Hilbert space has as a basis the list of fermionic elementary particles and the operator is the Yukawa coupling matrix. This worked fine for the minimal Standard Model,[†] but there was a problem of doubling the number of fermions, and also the Kamiokande experiments on solar neutrinos showed around 1998 that, because of neutrino oscillations, one needed a modification of the Standard Model incorporating in the leptonic sector of the model the same type of mixing matrix already present in the quark sector. One further needed to incorporate a subtle mechanism, called the see-saw mechanism, that could explain why the observed masses of the neutrinos would be so small. At first my reaction to this modification of the Standard Model was that it would certainly not fit with the noncommutative geometry framework and hence that the previous agreement with noncommutative geometry was a mere coincidence. I kept good relations with my thesis adviser Jacques Dixmier and he kept asking me to give it a try, in spite of my 'a priori' pessimistic view. He was right and after about eight years I realised[‡] that the only needed change (besides incorporating a right-handed neutrino per generation) was to make a very simple change of sign in the grading for the antiparticle sector of the model.[§] This not only delivered naturally the neutrino mixing, but also gave the see-saw mechanism and settled the above fermion doubling problem. The main new feature that emerges is that when looking at the above table of signs giving the KO-dimension, one finds

[†] Chamsedinne and Connes (1996).

[‡] Cf. Connes (2006).

[§] This was also done independently in Barrett (2007).

that the finite noncommutative geometry F is now of dimension 6 modulo 8! Of course the space F being finite, its metric dimension is 0 and its inverse line-element is bounded. In fact this is not the first time that spaces of this nature – i.e. whose metric dimension is not the same as the KO-dimension – appear in noncommutative geometry. This phenomenon had already appeared for quantum groups.

Besides yielding the Standard Model with neutrino mixing and making testable predictions (as we shall see in Section 4.8), this allowed one to hope that, instead of taking the finite geometry F from experiment, one should in fact be able to derive it from first principles. The main intrinsic reason for crossing by a finite geometry F has to do with the value of the dimension of spacetime modulo 8. We would like this KO-dimension to be 2 modulo 8 (or equivalently 10) to define the fermionic action, since this eliminates[†] the doubling of fermions in the Euclidean framework. In other words the need for crossing by F is to shift the KO-dimension from 4 to 2 (modulo 8).

This suggested to classify the simplest possibilities for the finite geometry F of KO-dimension 6 (modulo 8) with the hope that the finite geometry F corresponding to the Standard Model would be one of the simplest and most natural ones. This was recently done in our joint work with Ali Chamseddine.[‡]

From the mathematical standpoint our road to F is through the following steps:

(i) We classify the irreducible triplets $(\mathcal{A}, \mathcal{H}, J)$.
(ii) We study the $\mathbb{Z}/2$-gradings γ on $\mathcal{H}$.
(iii) We classify the subalgebras $\mathcal{A}_F \subset \mathcal{A}$ which allow for an operator D that does not commute with the centre of $\mathcal{A}$ but fulfils the 'order one' condition:

$$[[D, a], b^0] = 0 \qquad \forall\, a, b \in \mathcal{A}_F.$$

[†] Because this allows one to use the antisymmetric bilinear form $\langle J\xi, D\eta\rangle$ (for $\xi, \eta \in \mathcal{H}, \gamma\xi = \xi, \gamma\eta = \eta$).

[‡] Chamsedinne and Connes (2007).

The classification in the first step (i) shows that the solutions fall in two classes, in the first the dimension n of the Hilbert space $\mathcal{H}$ is a square: $n = k^2$, in the second case it is of the form $n = 2k^2$. In the first case the solution is given by a real form of the algebra $M_k(\mathbb{C})$ of $k \times k$ complex matrices. The representation is given by the action by left multiplication on $\mathcal{H} = M_k(\mathbb{C})$, and the isometry J is given by $x \in M_k(\mathbb{C}) \mapsto J(x) = x^*$. In the second case the algebra is a real form of the sum $M_k(\mathbb{C}) \oplus M_k(\mathbb{C})$ of two copies of $M_k(\mathbb{C})$ and while the action is still given by left multiplication on $\mathcal{H} = M_k(\mathbb{C}) \oplus M_k(\mathbb{C})$, the operator J is given by $J(x, y) = (y^*, x^*)$.

The study (ii) of the $\mathbb{Z}/2$-gradings shows that the commutation relation $J\gamma = -\gamma J$ excludes the first case. We are thus left only with the second case and we obtain among the very few choices of lowest dimension the case $\mathcal{A} = M_2(\mathbb{H}) \oplus M_4(\mathbb{C})$ where $\mathbb{H}$ is the skew field of quaternions. At a more invariant level the Hilbert space is then of the form $\mathcal{H} = \mathrm{Hom}_{\mathbb{C}}(V, W) \oplus \mathrm{Hom}_{\mathbb{C}}(W, V)$ where V is a 4-dimensional complex vector space, and W a 2-dimensional graded right vector space over $\mathbb{H}$. The left action of $\mathcal{A} = \mathrm{End}_{\mathbb{H}}(W) \oplus \mathrm{End}_{\mathbb{C}}(V)$ is then clear and its grading as well as the grading of $\mathcal{H}$ come from the grading of W.

Our main result then is that there exists up to isomorphism a unique involutive subalgebra of maximal dimension $\mathcal{A}_F$ of $\mathcal{A}^{\mathrm{ev}}$, the even part[†] of the algebra $\mathcal{A}$, which solves (iii). This involutive algebra $\mathcal{A}_F$ is isomorphic to $\mathbb{C} \oplus \mathbb{H} \oplus M_3(\mathbb{C})$ and together with its representation in $(\mathcal{H}, J, \gamma)$ gives the noncommutative geometry F which we used in Chamseddine, Connes and Marcolli (2006) to recover the Standard Model coupled to gravity using the spectral action which we now describe.

4.7 OBSERVABLES IN GRAVITY AND THE SPECTRAL ACTION

The missing ingredient, in the above description of the Standard Model coupled to gravity, is provided by a simple action principle – the

[†] One restricts to the even part to obtain an ungraded algebra.

spectral action principle† – that has the geometric meaning of 'pure gravity' and delivers the action functional (4.1) when evaluated on $M \times F$. To this action principle we want to apply the criterion of *simplicity* rather than that of *beauty* given the relative nature of the latter. Thus we imagine trying to explain this action principle to a Neanderthal man. The spectral action principle, described below, passes the 'Neanderthal test', since it amounts to counting spectral lines.

The starting point at the conceptual level is the discussion of observables in gravity. By the principle of gauge invariance the only quantities which have a chance to be observable in gravity are those which are invariant under the group of diffeomorphisms of the spacetime M. Assuming first that we deal with a classical manifold (and 'Wick rotate' to Euclidean spacetime signature for simplicity), one can form a number of such invariants (under suitable convergence conditions) as integrals of the form

$$\int_M F(K)\sqrt{g}\,\mathrm{d}^4x \tag{4.8}$$

where $F(K)$ is a scalar invariant function‡ of the Riemann curvature K. There are§ other more complicated examples of such invariants, where those of the form (4.8) appear as the *single integral* observables i.e. those which add up when evaluated on the direct sum of geometric spaces. Now while in theory a quantity like (4.8) is observable, it is almost impossible to evaluate since it involves the knowledge of the entire spacetime and is in that way highly nonlocalised. On the other hand, spectral data are available in localised form anywhere, and are (asymptotically) of the form (4.8) when they are of the additive form

$$\mathrm{Trace}\,(\,f(D/\Lambda)), \tag{4.9}$$

where D is the Dirac operator and f is a positive even function of the real variable. The parameter Λ fixes the mass scale. The spectral action

† Chamseddine and Connes (1996–2006).

‡ The scalar curvature is one example of such a function but there are many others.

§ See Giddings, Hartle and Marolf (2005).

principle asserts that the fundamental action functional S that allows to compare different geometric spaces at the classical level and is used in the functional integration to go to the quantum level, is itself of the form (4.9). The detailed form of the function f is largely irrelevant since the spectral action (4.9) can be expanded in decreasing powers of the scale Λ and the function f only appears through the scalars

$$f_k = \int_0^\infty f(v)\, v^{k-1}\, dv. \tag{4.10}$$

As explained above, the gauge potentials make good sense in the framework of noncommutative geometry and come from the inner fluctuations of the metric.

Let M be a Riemannian spin 4-manifold and F the finite noncommutative geometry of KO-dimension 6 described above. Let $M \times F$ be endowed with the product metric. Then†

(i) The unimodular subgroup of the unitary group acting by the adjoint representation $\mathrm{Ad}(u)$ in $\mathcal{H}$ is the group of gauge transformations of the Standard Model.

(ii) The unimodular inner fluctuations of the metric give the gauge bosons of the Standard Model.

(iii) The full Standard Model (with neutrino mixing and see-saw mechanism) minimally coupled to Einstein gravity is given in Euclidean form by the action functional

$$S = \mathrm{Tr}(f(D_A/\Lambda)) + \frac{1}{2}\langle J\,\tilde{\xi}, D_A\,\tilde{\xi}\rangle, \quad \tilde{\xi} \in \mathcal{H}^+_{cl},$$

where D_A is the Dirac operator with the unimodular inner fluctuations.

The change of variables from the Standard Model to the spectral model is summarised in Table 4.1.

4.8 PREDICTIONS

The above spectral model can be used to make predictions assuming the 'big-desert hypothesis' (absence of new physics up to unification

† Chamseddine, Connes and Marcolli (2006).

Table 4.1 *Conversion from Standard Model to the Spectral Action, cf. Chamseddine, Connes and Marcolli (2006).*

Standard Model	Spectral Action
Higgs boson	Inner metric$^{(0,1)}$
Gauge bosons	Inner metric$^{(1,0)}$
Fermion masses u, ν	Dirac$^{(0,1)}$ in $\uparrow$
CKM matrix masses down	Dirac$^{(0,1)}$ in $(\downarrow 3)$
Lepton mixing masses leptons e	Dirac$^{(0,1)}$ in $(\downarrow 1)$
Majorana mass matrix	Dirac$^{(0,1)}$ on $E_R \oplus J_F E_R$
Gauge couplings	Fixed at unification
Higgs scattering parameter	Fixed at unification
Tadpole constant	$-\mu_0^2 \, \lvert\mathbf{H}\rvert^2$
Graviton	Dirac$^{(1,0)}$

scale) together with the validity of the spectral action as an effective action at unification scale. While the big-desert hypothesis is totally improbable, a rough agreement with experiment would be a good indication for the spectral model.

When a physical theory is described at the classical level by an action principle i.e. by minimising an action functional S on the configurations, there is a heuristic prescription due to R. Feynman in order to go from the classical to the quantum. This prescription affects each classical field configuration with the probability amplitude:

$$e^{\iota\, S/\hbar}.$$

The prescription is only heuristic for a number of reasons, one being that the overall sum over all configurations is an oscillatory integral whose convergence is hard to control. This situation improves if one works in the Euclidean formulation. This means that our configurations are Euclidean and in the functional integral the weight of such

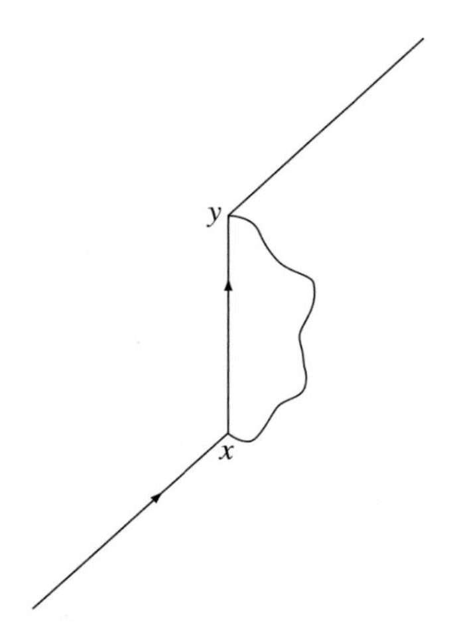

FIG. 4.12 The arrow from x to y is the propagator for fermions.

a configuration is now given by

$$e^{-S/\hbar}.$$

After passing to the Euclidean spacetime formulation, one can start computing the Feynman integral applying perturbative techniques, but one quickly meets a basic problem due to the omnipresence of divergent integrals in evaluating the contribution of simple processes (called Feynman graphs) such as the emission and absorption of a photon by an electron shown in Figure 4.12. This problem was already present before the Feynman path integral formulation, when trying to compute the higher-order terms in Dirac's theory of absorption and emission of radiation. Physicists discovered around 1947 a procedure, called renormalisation, to handle this problem. The physics idea is to make the distinction between the so-called bare parameters which enter in the mathematical formula for the action S and the observed parameters such as masses, charges etc. This distinction goes back to the work of Green in hydrodynamics in the nineteenth century. It is easy to explain in a simple example: the motion of a ping-pong ball inside water (Figure 4.13). What Green found is that Newton's law $F = ma$ remains valid, but the mass m is no longer the bare mass m_0 that would be obtained by weighing

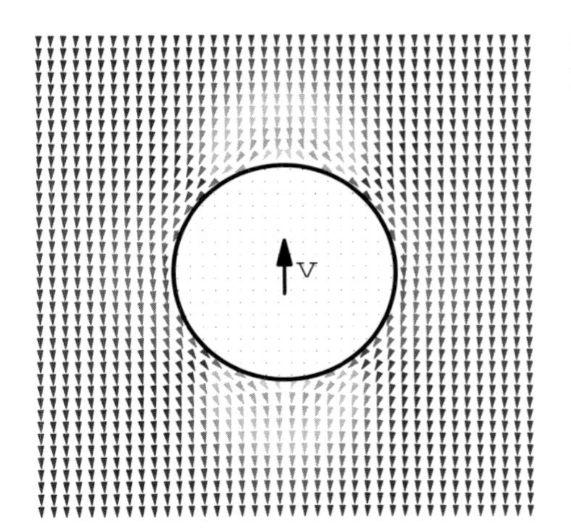

FIG. 4.13 Hydrodynamics of a ping-pong ball, cf. Green 1830.

the ping-pong ball with a scale outside the water, but a modified mass

$$m = m_0 + \frac{1}{2}M$$

where M is the mass of the same volume of water. In other words, due to the presence of the surrounding field of water the inertial mass of the ball is increased, as if it were half full of water! This 'renormalisation' of mass has direct measurable effects even on the initial acceleration of the ping-pong ball shooting back to the surface (it is about seven times smaller than what one would compute using the Archimedean law).

In order to take care of the divergent integrals that are omnipresent in the perturbative calculations of quantum field theory, one introduces an energy scale Λ (called a 'cutoff' scale) and one only restricts the integration variable to values of the energy which are smaller than Λ (this is the 'cutoff'). One then cleverly lets the 'constants' which enter in the formula for the action S depend on Λ so that the infinities disappear (in the computations of observable quantities) when one removes the cutoff, i.e. when Λ goes to infinity. This magic way of sweeping the difficulty under the rug works for theories which are 'renormalisable' and gives in electrodynamics incredibly precise

FIG. 4.14 Évariste Galois.

predictions which, for the anomalous moment of the electron, are in perfect agreement with the observed value. The price one pays is that this variability of the constants (they now depend on Λ) introduces a fundamental ambiguity in the renormalisation process.

The corresponding symmetry group is called the renormalisation group, and in a recent work with M. Marcolli we exhibited an incarnation of this group as a universal symmetry group – called the 'cosmic Galois group' following P. Cartier – of all renormalisable theories. As the name suggests it is deeply related to the ideas of Galois which he briefly sketched in his last letter:

> *Tu sais, mon cher Auguste, que ces sujets ne sont pas les seuls que j'aie explorés. Mes principales méditations depuis quelque temps étaient dirigées sur l'application à l'analyse transcendante de la théorie de l'ambiguïté. Il s'agissait de voir a priori dans une relation entre des quantités ou fonctions transcendantes quels échanges on pouvait faire, quelles quantités on pouvait substituer aux quantités données sans que la relation pût cesser d'avoir lieu. Cela fait reconnaitre tout de suite l'impossibilité de beaucoup*

> *d'expressions que l'on pourrait chercher. Mais je n'ai pas le temps et mes idées ne sont pas encore assez développées sur ce terrain qui est immense.*†

It might seem that the presence of the above ambiguity precludes the possibility to predict the values of these physical constants, which are in fact not constant but depend upon the energy scale Λ. In fact the renormalisation group gives differential equations which govern their dependence upon Λ. The intuitive idea behind this equation is that one can move down i.e. lower the value of Λ to $\Lambda - d\Lambda$ by integrating over the modes of vibrations which have their frequency in the given interval. For the three coupling constants g_i (or rather their squares α_i) which govern the three forces (excluding gravity) of the Standard Model, their dependence upon the scale is shown in Figure 4.15 and shows that they are quite different at low scale, but they become comparable at scales of the order of 10^{15} GeV. This suggested long ago the idea that physics might become simpler and 'unified' at scales (called unification scales) of that order.

We can now describe the predictions obtained by comparing the spectral model with the Standard Model coupled to gravity. The status of 'predictions' in the above spectral model is based on two hypotheses:

(i) The model holds at unification scale.
(ii) One neglects the new physics up to unification scale.

The value of that scale is the above unification scale since the spectral action delivers the same equality $g_3^2 = g_2^2 = \frac{5}{3}\, g_1^2$, which is

† You know, dear Auguste, that these topics are not the only ones I explored. My main thoughts in recent times were directed towards applying the theory of ambiguity to transcendental analysis. The issue is to decide 'a priori' in a relation between transcendental quantities or functions, which exchanges can be made, which substitutions of new quantities instead of the given ones can be performed without altering the relation. This method allows one to recognise immediately the impossibility of many expressions one could guess. But I do not have the time and my ideas are not sufficiently developed in this immense field.

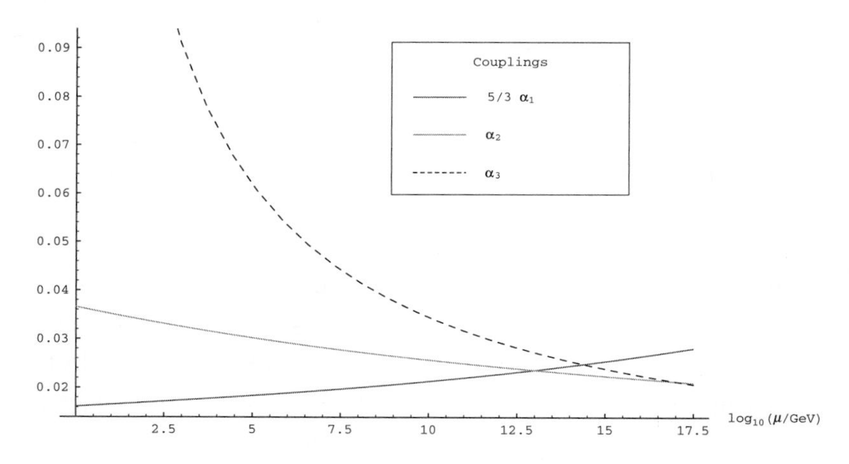

FIG. 4.15 The running of the three couplings. The scale is called μ rather than Λ in the plot and is expressed in a logarithmic scale with base 10 and in GeV.

common to all 'Grand-Unified' theories. It gives more precisely the following unification of the three gauge couplings:

$$\frac{g_3^2\, f_0}{2\pi^2} = \frac{1}{4}, \qquad g_3^2 = g_2^2 = \frac{5}{3}\, g_1^2 \,.$$

Here $f_0 = f(0)$ is the value of the test function f at 0.

The second feature which is predicted by the spectral model is that one has a see-saw mechanism for neutrino masses with large $M_R \sim \Lambda$.

The third prediction that one gets by making the conversion from the spectral model back to the Standard Model is that the mass matrices satisfy the following constraint at unification scale:

$$\sum_\sigma (m_\nu^\sigma)^2 + (m_e^\sigma)^2 + 3\,(m_u^\sigma)^2 + 3\,(m_d^\sigma)^2 = 8\, M_W^2.$$

In fact it is better to formulate this relation using the following quadratic form in the Yukawa couplings:

$$Y_2 = \sum_\sigma (y_\nu^\sigma)^2 + (y_e^\sigma)^2 + 3\,(y_u^\sigma)^2 + 3\,(y_d^\sigma)^2$$

so that the above prediction means that

$$Y_2(S) = 4\, g_3^2.$$

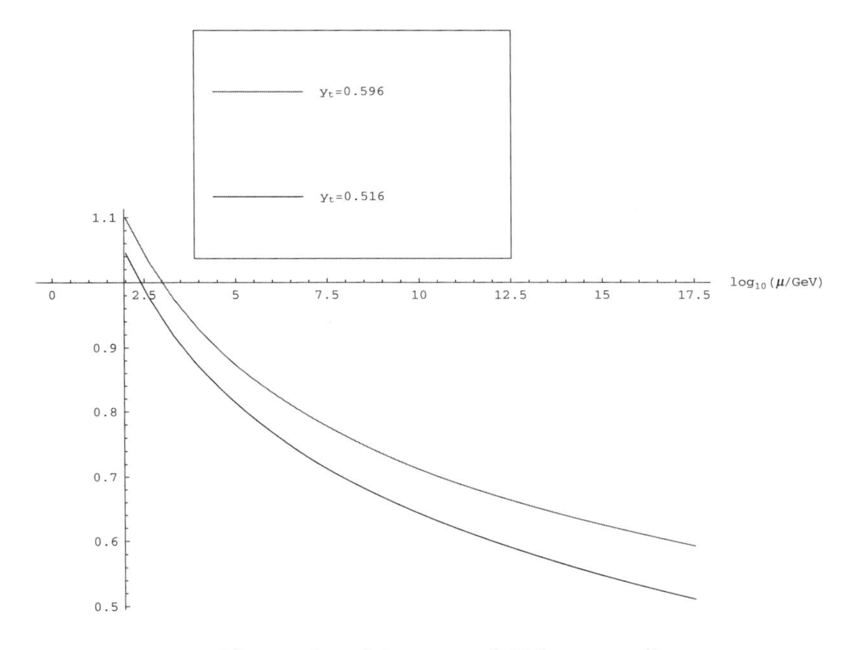

FIG. 4.16 The running of the top quark Yukawa coupling.

This, using the renormalisation group (cf. Figure 4.16) to compute the effective value at our scale, yields a value of the top-quark mass which is 1.04 times the observed value when neglecting the Yukawa couplings of the bottom quarks etc. and is hence compatible with experiment.

Another prediction obtained from the conversion table is the value of the Higgs scattering parameter at the unification scale:

$$\tilde{\lambda}(\Lambda) = g_3^2 \, \frac{b}{a^2}.$$

The numerical solution to the renormalisation group equations (cf. Figure 4.17) with the boundary value $\lambda_0 = 0.356$ at $\Lambda = 10^{17}$ GeV gives $\lambda(M_Z) \sim 0.241$ and a Higgs mass of the order of 168 GeV.

Finally since our theory unifies the Standard Model with gravity it also predicts the value of the Newton constant at unification scale. Here one needs to be really careful since, while renormalisation works

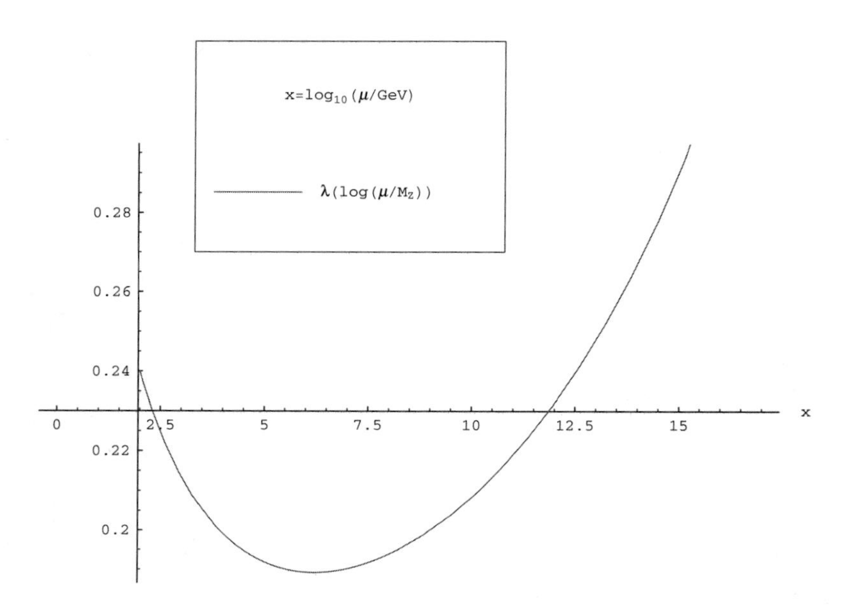

FIG. 4.17 The running of the quartic Higgs coupling.

remarkably well for the quantisation of the classical fields involved in the Standard Model, this latter perturbative technique fails when one tries to deal with the gravitational field $g_{\mu\nu}$ using the Einstein–Hilbert action. However the spectral action delivers additional terms like the square of the Weyl curvature, and one can then use this action as an effective action. This means that one does not pretend to have a fundamental theory i.e. a theory that will run at any energy, but rather an action functional which is valid at a certain energy scale. Even then, the running of the Newton constant under the renormalisation group is scheme dependent and not well understood. However making the simple hypothesis that the Newton constant does not change much up to a unification scale of the order of 10^{17} GeV one finds that it suffices to keep a ratio of order one between the moments† of the test function f to obtain a sensible value. We also checked that the additional terms such as the term in Weyl curvature square, do not

† Namely $f_2 \sim 5 f_0$ where f_2 is defined by (4.10).

have any observable consequence at ordinary scales (the running of these terms is known and scheme independent).

4.9 WHAT ABOUT QUANTUM GRAVITY?

As explained above, the theory which is obtained from the spectral action applied to noncommutative geometries of the form $M \times F$ is not a fundamental theory but rather an effective theory, in that it stops making sense above the unification scale. For instance the gravitational propagator admits a tachyon pole and unitarity breaks down. It is thus natural to wonder how such a theory can emerge from a more fundamental one. In other words one would like to get some idea of what happens at energies above the unification scale, if that has any meaning.

In our book Connes and Marcolli (2007) we have developed an analogy between the role of spontaneous symmetry breaking in a number-theoretic framework – intimately related to the spectral realisation of the zeros of the Riemann zeta function – and the symmetry breaking in the electro-weak sector of the Standard Model.

It raises in particular the possibility that geometry only emerges after a suitable symmetry-breaking mechanism which extends to the full gravitational sector the electro-weak symmetry breaking that we discussed above. The invariance of the spectral action under the symplectic unitary group in Hilbert space is broken during this process to the compact group of isometries of a given geometry.

Thus, if one follows closely the analogy with the number-theoretic system, one finds that above a certain energy (the Planck energy, say) the very idea of a spacetime disappears and the state of the global system is a mixed state of type III (in the sense that it generates an operator algebra factor of type III) whose support consists of operator theoretic data with essentially no classical geometric meaning. In the number-theoretic system the role of the energy level is played by temperature, and the key notion is that of KMS state. Thus if one follows this analogy one needs to abandon the idea of the initial singularity of spacetime and replace it by the emergence of geometry, through a symmetry-breaking phenomenon. In particular, the idea of trying to quantise the gravitational field in a fixed background spacetime manifold by finding a renormalisable unitary quantum field theory becomes irrealistic.

BIBLIOGRAPHY

Adler, K. (2002) *The Measure of All Things: The Seven-Year Odyssey that Transformed the World*, Free Press.

Barrett, J. (2007). The Lorentzian version of the noncommutative geometry model of particle physics, *J. Math. Phys.* **48**, 012303.

Chamseddine, A. and Connes, A. (1996) Universal formula for noncommutative geometry actions: unification of gravity and the Standard Model, *Phys. Rev. Lett.* **77**, 4868–71.

Chamseddine, A. and Connes, A. (1997) The spectral action principle, *Comm. Math. Phys.* **186**, 731–50.

Chamseddine, A. and Connes, A. (2005) Scale invariance in the spectral action, hep-th/0512169 to appear in *J. Math. Phys.*

Chamseddine, A. and Connes, A. (2006) Inner fluctuations of the spectral action, hep-th/0605011.

Chamseddine, A. and Connes, A. (2007) Why the Standard Model. arXiv:0706.3688; A dress for SM the beggar. arXiv:0706.3690.

Chamseddine, A., Connes, A. and Marcolli, M. (2006) Gravity and the standard model with neutrino mixing, hep-th/0610241.

Connes, A. (1994) *Noncommutative Geometry*, Academic Press.

Connes, A. (1995) Geometry from the spectral point of view, *Lett. Math. Phys.* **34**, no. 3, 203–38.

Connes, A. (1996) Gravity coupled with matter and the foundation of noncommutative geometry, *Comm. Math. Phys.* **182**, no. 1, 155–76.

Connes, A. (2006) Noncommutative geometry and the standard model with neutrino mixing, *JHEP* 0611:08.

Connes, A. and Dubois-Violette, M. (2002) Noncommutative finite-dimensional manifolds. I. Spherical manifolds and related examples, *Comm. Math. Phys.* **230**, no. 3, 539–79.

Connes, A. and Dubois-Violette, M. (2003) Moduli space and structure of noncommutative 3-spheres, *Lett. Math. Phys.* **66**, no. 1-2, 91–121.

Connes, A. and Dubois-Violette, M. (2005) Noncommutative finite-dimensional manifolds II. Moduli space and structure of noncommutative 3-spheres. ArXiv math.QA/0511337.

Connes, A. and Landi, G. (2001) Noncommutative manifolds, the instanton algebra and isospectral deformations, *Comm. Math. Phys.* **221**, no. 1, 141–59.

Connes, A. and Lott, J. (1991) Particle models and noncommutative geometry, in *Recent Advances in Field Theory (Annecy-le-Vieux, 1990), Nuclear Phys. B Proc. Suppl.* **18B**, 29–47.

Connes, A. and Marcolli, M. (2007) *Noncommutative Geometry, Quantum Fields and Motives*, in press, American Mathematical Society and Hindustan Book Agency.

Giddings, S., Hartle, J. and Marolf, D. (2005) Observables in effective gravity, hep-th/0512200.

Heisenberg, W. (1925) Über quantentheoretische Umdeutung kinematischer und mechanischer Beziehungen, *Z. Phys.* **33**, 879–93.

Kastler, D. (1993) A detailed account of Alain Connes' version of the standard model in noncommutative geometry. I, II, *Rev. Math. Phys.* **5**, no. 3, 477–532.

Kastler, D. (1995) The Dirac operator and gravitation, *Commun. Math. Phys.* **166**, 633-43.

Kastler, D. (2000) Noncommutative geometry and fundamental physical interactions: the Lagrangian level, *J. Math. Phys.* **41**, 3867–91.

Sullivan, D. (2005) *Geometric Topology: localization, periodicity and Galois symmetry*, The 1970 MIT notes, edited and with a preface by Andrew Ranicki, K-Monographs in Mathematics, vol 8, Springer.

van der Waerden, B. L. (1967) *Sources of Quantum Mechanics*, Dover.

5 Where physics meets metaphysics

Michael Heller

5.1 INTRODUCTION

Physics and metaphysics are two distinct occupations of human beings, and not long ago a lot of effort was invested into keeping them strictly apart. Nowadays, however, the situation seems to be changing. A couple of years ago, an international conference was organised at Cambridge, UK, the proceedings of which bear the title 'Physics Meets Philosophy at the Planck Scale'.[†] What is special about the Planck Scale that physics and philosophy (of which metaphysics is an essential part) seem to have something to tell to each other?

In physicists' jargon the 'Planck Scale' or the 'Planck era' means either the most fundamental level of the physical Universe, or an edge at which our present theories of physics break down (this is why the Planck era is also called the Planck threshold). Currently, these two meanings are almost synonymous since the fundamental theory of physics lies beyond the reach of our well-founded physical theories and models.

There are two directions along which we could approach the Planck era. We can either adopt the path followed by cosmologists, or that followed by elementary particle physicists. In cosmology, one tries to reconstruct the history of the Universe starting from our present era as far backward in time as possible. As one moves in this direction, the Universe contracts and becomes denser and denser, till one reaches a density of the order of 10^{95} g/cm^3, and then one finds

[†] Callender and Huggett (2001).

On Space and Time, ed. Shahn Majid. Published by Cambridge University Press.

oneself at the Planck era. In elementary particle physics, one tries to obtain higher and higher energies, with the help of which one can penetrate smaller and smaller distances. If one were able to penetrate a distance of the order of 10^{-33} cm (which is far beyond the present possibilities), one would find oneself at the Planck era. Here we encounter the first philosophically striking property. Irrespective of which of the above two directions you choose, you reach *the same* Planck era. There are not two Planck eras, one in a distant past and another now on the deepest space level; it is the same threshold of our knowledge. It looks as if space and time behaved in an unusual way on approaching the fundamental level.

Moreover, Einstein's theory of General Relativity predicts (or rather retrodicts) that the present evolution of the Universe started with a state, called initial singularity, with enormously high densities and pressures. Let us suppose that we begin counting cosmic time at the initial singularity, i.e. the initial singularity occurs at the moment $t = 0$. Then, theoretically speaking, if time goes to zero, the density of the Universe goes to infinity. This means that the Planck threshold, characterised by a density of the order of 10^{95} g/cm^3, is somewhere in the vicinity of the initial singularity. Indeed, some compatibility arguments require us to assume that the Planck threshold occurred when the cosmic clock indicated $t = 10^{-43}$ s. The point is, however, that at that moment our present theories break down, together with the general theory of relativity on which the prediction of the initial singularity is based and, strictly speaking, we do not know whether the initial singularity actually occurred, or not. To make things worse, in the popular literature and in the media, the term 'Big Bang' is used to denote either the initial singularity, or the 'state with enormous densities and pressures, out of which everything began', or both indiscriminately.

It goes without saying that all boundaries of scientific research are but a challenge to scientists. The Planck threshold is not an exception. There exist many attempts at discovering the physics of the Planck era. Two major physical theories shape the structure of the present Universe: Einstein's theory of gravity, called the general theory of relativity, and quantum mechanics. The former rules at the scales of stars, galaxies and clusters of galaxies, the latter at the scale of elementary particles and quantum fields. There are strong reasons to think that on the fundamental level both these theories are unified, and only when crossing the Planck threshold does this unified physics split into the present-day somewhat schizophrenic world of two separated theories. The great challenge for contemporary theoretical physics is to create a quantum-gravity theory, a unification of gravity and quanta, as a necessary condition to penetrate beyond the Planck threshold.

The place at which our reasonably well-founded theories border with the unknown always constitutes philosophical provocation. Quite often we prefer to invent pseudo-explanations than to admit that we do not know. And even if this is not the case, our imagination suggests various possibilities, some of which seem to be worthwhile to be considered and reconsidered. In this way, physics provokes metaphysics. If this happens in the field of cosmology, metaphysics often assumes the form of theology. To explain the Universe it seems unavoidable to go beyond it. Or, is it possible to explain the Universe in terms of the Universe itself? When stating this dilemma, we are already in the realm of natural theology (irrespectively of the option we prefer). The aim of this essay is to present my own views on these matters. Being based on traditional Christian theology, they do not pretend to be original, but if we remember that theology is, roughly speaking, a rational reflection on the doctrinal content of a given religion, it is obvious that good theology should adapt itself to new data that enrich our general knowledge. These data can come from various sources, among others from social conditions, from the cultural environment, and also from science. The latter source is especially important for theological reflection on the physical world. As I have just mentioned, my views in this respect do not pretend to originality. Their value (if they have any) follows from my scientific background. Working in the field of cosmology and theoretical physics,

I feel obliged to somehow coordinate my religious belief with my scientific insights. The matter is delicate. It was not without sound reasons that science and theology were kept separated. So-called 'building bridges' between science and theology without very balanced methodological care easily results in doctrinal anarchy, and even deepens the existing conflict between them. This is not to say that there should be no interactions between these two areas of human activity. They always were in history and, as we can reasonably guess, they always will be, irrespectively of any methodological barriers constructed by purists on both sides. The point is, however, that the interface between science and theology must be based on extremely 'fine-tuned' principles. Any deviation from the right track leads exponentially to a catastrophe. In the following sections, I shall be audacious enough to balance on this dangerous borderline, hoping only that the exponential precipice will not engulf me and my readers (in the last section I shall make some clarifications in an attempt not to trespass against the principle of methodological honesty). To ensure this, let us begin with a few theological premises that are vitally important as far as my main topic is concerned.

5.2 FALLACY OF GOD IN THE GAPS

For a believing Christian there is a natural temptation to see in the Big Bang the act of creation of the Universe by God. Many theologians, philosophers, and also some scientists support this view. It seems natural since if the Big Bang is indeed the beginning of everything that exists, the initiation of existence could, it seems, be attributed only to the creative power of God. However, one should be cautious in formulating such claims. As we have seen above, the idea of the Big Bang is, in fact, a cover for a gap in our knowledge. Before we go back to the initial moment, we must meet the Planck threshold, an edge at which our present physical theories break down. Although we do have some preliminary versions of the theory of quantum gravity, they are rather far away from anything close to its final form, and as long as we do not have it, we shall not know whether the initial singularity actually existed, or not. Therefore, if we persist in identifying the Big Bang with God's act of creation, we are basing our theological doctrine on a gap in our knowledge. And the 'God of the gaps' strategy has a bad name in theology.

Although the idea itself arose much earlier, the term 'God of the gaps' goes back to Newton's time. By using Newtonian laws of motion it was possible to compute trajectories of planets with great preciseness, but the initial conditions of planetary motions were not determined by mechanical laws. Newton believed that there exists an Intelligent Clock Master who predetermined initial conditions and gave the initial momentum to the mechanism of the Universe. Moreover, He has to intervene from time to time to reestablish the original harmony when external influences (e.g. caused by visits of comets) disturb the structure of the planetary system. Later on, when classical mechanics made enormous progress, God's interventions into the course of heavenly motions turned out to be unnecessary, and the 'hypothesis of God' seemed superfluous. Similar processes occurred in other branches of science, for instance in biology, where the examples of apparent finality, or 'intelligent design', were gradually replaced by natural selection and genetic mutations. In consequence, the strategy of filling in gaps in our knowledge with the 'hypothesis of God' started to be viewed by the majority of theologians as a compromised doctrine. One should look for God not where our knowledge breaks down, but rather where it is solid and well founded.

We should learn this lesson from history. When thinking about 'science and theology', our first methodological principle is that we should not fall into the trap of the 'God of the gaps' strategy. This principle is purely negative; it says what should be avoided. The second principle, at least as far as my personal choice is concerned, is positive (it is doctrinal rather than methodological); it indicates a direction which my interpretative efforts should follow. It says that all my analyses and interpretations should be carried out within the conceptual space delineated by a Christian doctrine of creation.

5.3 CHRISTIAN DOCTRINES OF CREATION

In traditional theology, the unique relationship between the world and God, consisting in the total and continuous dependence of the world in its existence on God, is called *creation*. If, at a certain instant, God had broken this dependence, the world would have turned into nothingness. In this sense one should understand the popular saying that 'to create is to make something out of nothing'. Let us notice that in the above formulation there is no mention about temporal beginning. The concept of creation is an answer to the famous question of Leibniz 'Why is there something rather than nothing?' This question better than anything else encapsulates the metaphysical weight of the creation problem, and emphasises its strong global flavour. By asking this question we restrict the problem of the world's existence neither to any of its particular properties, nor to any particular place or time. The question of time does not even arise.

The question of time and creation has a rich history. It goes back to St Augustine and a famous question from the Eleventh Book of his *Confessions*: 'What was God doing before He made heaven and earth?' After dismissing the answer: 'He was preparing hell for those who pry into mysteries', as 'facetious' and 'avoiding the pressure of the question', Augustine gives his own answer dressed in the form of a prayer: 'But if before heaven and earth there was no time, why is it asked, What didst Thou then? For there was no 'then' when time was not'.[†] Creation of the world is not a temporal act. The world was not created *in time*, the world is created *with time*.

Another important step in this direction was made by St Thomas Aquinas. In the thirteenth century the Christian West recovered, through Arabs, important writings of Aristotle. Soon it became obvious that his natural philosophy was superior with respect to everything Western thinkers had known about it thus far. But Aristotle taught that the Universe existed always, it had no beginning. It seemed that this doctrine was incompatible with the Christian idea

[†] *Confessions*, Book XI, 12 and 13.

FIG. 5.1 Bennozo Gozzoli's 1465 fresco 'Take up and Read' from his cycle on the life of St Augustine of Hippo 354–430.

of creation *ex nihilo*. This was one of the main reasons of several condemnations of Aristotelian teaching, the most famous being the condemnation of 213 theses, believed to be Aristotelian, by Étienne Tempier, Bishop of Paris. It was Aquinas who attempted to show that there is no contradiction between Aristotelian science and the Christian doctrine. The key to his analysis is the distinction between *change* that is involved in time, and *creation* that is essentially atemporal. Creation has nothing to do with producing a change in something (which would presuppose pre-existing material), but is the radical causing of existence and, as such it occurs always (from the point of view of creaturely things). As a consequence, Aquinas saw no contradiction between the concept of the Universe existing from 'temporal minus infinity' (with no beginning) and the concept of creation *ex nihilo*. For, even if the Universe had no beginning in time, it still would depend on God as the Primary Cause of its existence. The possibility, very seriously considered by St Thomas Aquinas, that the world could exist from 'temporal minus infinity' and nevertheless be created by God, emphasises an atemporal aspect of the idea of creation.

A widely spread opinion that the creation of the Universe presupposes its temporal beginning is the consequence of the enormous success of Newtonian physics. It was Newton who claimed that everything must exist in space and time, but space and time themselves are independent of anything else. In this sense, space and time are absolute. Newton believed that before the creation of the world there had been 'empty space' and 'empty time', and that the act of creation (understood as the initiation of the world's existence) had occurred at a certain moment of absolute time. Owing to great successes of classical physics, these views have been incorporated into the image of the world accepted by our culture. Even the majority of people who reject the idea of creation reject it in its Newtonian formulation.

However, many contemporary theologians prefer to adhere to the traditional doctrine. One of them writes that a 'Being from whom the existence of all things derives' cannot be subject to any temporal constraints: 'Time is a condition of the creature, a sign of dependence. ... The act of creation is a single one, in which what is past, present or future from the perspective of the creature issues as a single whole from the Creator.'[†]

The above presented doctrine of creation has nothing in common with the God of the gaps ideology. The act of creation does not fill in any hole in our knowledge; on the contrary, it is a kind of global explanation of the existence of everything that exists. It does not exclude the possibility that God indeed initiated the existence of the Universe in the initial singularity, but it does not presuppose it.

Moreover, creation is not only an explanation of the brute fact of the world's existence, but also an explanation of the world's existence in all its sophisticated and artful richness. All so far discovered laws of physics, and all not yet discovered laws of physics, are but elements or aspects of this richness. The creation of the Universe is a rational process, and it is this rationality that is the *raison d'être*

[†] McMullin (1997).

of science. It this context, Einstein's saying 'The only thing I want to know is the Mind of God he had in his act of creation' acquires its full meaning.

We thus have two general premises for our further considerations. One is negative – we should avoid any form of the God of the gaps ideology; the second is positive – we want to remain within the broad context of the doctrine of creation as it has been sketched above. However, in this doctrine some other philosophical concepts are involved such as, among others, the concepts of time, space and causality, and our understanding of these concepts matures together with the progress of science. They changed their meaning in going from classical physics to contemporary physics, and they will certainly change their meanings when a quantum-gravity theory is finally created. Our partial results in this area indicate the direction of these changes.

In the last three centuries encounters between science and theology often ended with clash and conflict. The aim of the present essay is to see whether science and theology can nowadays interact with profit for both sides. The profit for theology is obvious; the question at stake is to become relevant for men and women in our times. To see profit for science is perhaps less obvious, but we should take into account the fact that much of Western science, such as for instance Newton's ideas, are imbued with things taken ultimately from theology, and it is better to be aware of this influence than not. To understand science is a part of understanding the world. In Sections 5.4–5.8, we provide some of the scientific background, indispensable to pursue this goal, after which, in the final part of the essay (Sections 5.9–5.13), we turn to wider aspects of the discussion.

5.4 QUANTUM-GRAVITY CHALLENGE

The evolution of concepts is one of the principal forces driving scientific progress. For instance, the entire history of Einstein's theory of relativity can be reduced to a careful (and sometimes painful) analysis of concepts such as: space, time, inertia, mass, gravity, causality, distance, curvature. When a new theory inherits some concepts from an old theory, the concepts usually change their meanings, and sometimes transform into new ones. However, concepts do not evolve by themselves. They are always involved in solving problems. It often turns out that some concepts are too narrow to help in solving a posed problem. When they are enforced to do so, they produce inconsistencies and paradoxes. Such a crisis usually initiates a new conceptual revolution. There are no isolated concepts in science. Any change in the meaning of one concept ignites a chain reaction of shifts of meanings in many other concepts. We should speak about evolution within the conceptual framework rather than about the evolution of isolated concepts.

There are many patterns according to which concepts evolve, but usually they become more and more general. However, this process is not a linear one. It can have many side-branches and diverse ramifications. The objective of science is to explain each specific phenomenon, but to explain a phenomenon in science means to put it in a more general conceptual scheme. Moreover, in this generalisation process old concepts are not eliminated but engulfed by new ones as their 'special cases'. We could risk the statement that the generalisation of conceptual schemes determines an 'arrow of time' of scientific progress.

Science itself is not an enterprise isolated from other human activities; it strongly interacts with other areas of our culture, especially with philosophy and theology. Consequently, the evolution of concepts is not limited to physical theories or to science in general; it often embraces philosophy and theology as well. An interesting example of this phenomenon is what could be called 'migration of concepts' from one of these disciplines to another. Let us consider, for instance, such concepts as space, time, causality, determinism. They began their 'official life' (after being inherited from everyday language) in religious and philosophical contexts. From theological and philosophical domains they migrated to scientific theories, and sometimes, after being transformed by scientific transmutations, they went back into philosophical or theological speculations. During every such metamorphosis they changed their meanings by adapting themselves to a new environment. The history of ideas is full of instances of these processes. But it would be naïve to think that they belong only to the past.

As we have seen, one of the greatest challenges of current theoretical physics is to create a theory that would unify two great, but so far only weakly interacting, areas of physics: General Relativity and quantum mechanics. The root of the problem consists in the very different physical natures of gravity (studied by General Relativity) and of the quantum world. This difference is reflected in the mathematical structures that are used to model gravity and quanta. Physicists are currently looking for a sufficiently rich mathematical structure that, when suitably interpreted, would somehow contain in itself mathematical structures responsible for both gravitational and quantum phenomena. It is rather obvious that when this objective is finally reached, it will entail a greater conceptual revolution. It goes without saying that, sooner or later, this revolution will have great repercussions in both philosophy and theology. Although we are still far away from the final result, the search for the unifying theory is so advanced that at least some of its consequences can be foreseen with some confidence.

There are various approaches to the search for quantum gravity. Let us enumerate the most popular of them: superstring theory together with its recent generalisation called M-theory, quantum loop theory, twistor theory, various approaches based on noncommutative geometry and quantum group theory. The reader will find some of them in various pages of the present book. In the following, I am not interested in any particular approach, but rather in conceptual perspectives opened by the problem itself of penetrating the most fundamental level of physics. If below I use examples taken from models based on noncommutative geometry rather than from other approaches, it is because of my personal interest and the fact that noncommutative geometry offers extraordinarily rich conceptual possibilities, the significance of which goes far beyond the area of theoretical physics. Even if noncommutative models will finally find no application to quantum gravity, they could still be interesting from a more general point of view. Being mathematical models, they display a strict logical network of dependencies which often reveal an unexpected interaction of concepts so far regarded as essentially independent from each other. And exactly at this point a vast field opens of new possibilities in the area of philosophical and theological speculations. One such possibility seems to me especially pregnant with consequences, namely the idea that the fundamental level can be nonlocal, i.e., that on this level there can be no time and no space in their usual sense. In the following, I shall amply explore this possibility but, before we enter this area, we must embark on an excursion into the world of modern geometry.

5.5 NONLOCAL SPACES

Everyone remembers, from one's early adventures with mathematics, the following rules: if the same terms appear on both sides of an equation, they can be cancelled; and a term can be transferred to the other side of an equation provided its sign is changed to the opposite one. In the eleventh century, the Persian mathematician Muhammad ibn Mūsā al-Khwārizmī wrote a treatise *Al-Kitab al-Jaber wa-l-Muqabala* (A Book on Calculation by Completing and Balancing) in which he discussed the above properties of equations. The title of this treatise gave rise to the name 'al-gebra' for a branch of mathematics dealing with solving equations. In 1591, the French mathematician François Viète started to replace numerical coefficients by letters (as it was the custom of doing in geometry). The term 'algebra' gradually changed its meaning from solving equations to calculating with the help of letters, later on called 'symbols'. Calculation with the help of symbols instead of numbers turned out to be more general and more effective. When new mathematical objects were discovered, such as sets, vectors and matrices, one symbolic formula (expressing, for instance, mathematical operations of addition and multiplication) could refer to all of them. The next step was an investigation of the properties of mathematical operations with no regard to objects on which they are defined. In this way *abstract algebra* was born.

The above one-paragraph history of algebra nicely shows the general tendency of mathematical progress: from particular to general, from concrete to abstract. No wonder that if we look at developments of mathematics in the last decades, we easily see that one of its main trends is a tendency to 'algebraisation'. This tendency can be traced back to Descartes who showed us how to translate standard geometric reasoning into algebraic calculations. In modern differential geometry, the concept of geometric space has been rigorously elaborated and is now known under the technical term of *differential manifold* or simply *manifold*. The standard way of defining this concept is in terms of coordinate systems with which this 'space' can be covered. The idea is similar to the one we use in reading maps: geographical latitude and geographical longitude are but coordinates which together form a coordinate system or a coordinate patch. To cover a more extended part of the earthly globe we must have a few coordinate patches, and they must 'overlap smoothly' on their bordering regions. In other words, we must use a few maps forming an atlas. Exactly the same happens in differential geometry: the concept of a general differential manifold is defined in terms of *maps* and *atlases* (which now become technical concepts).

There is, however, another method of defining differential manifolds that was known for a long time. Instead of coordinate systems on a manifold we can consider a family of all smooth functions on this manifold. By smooth function here we mean an assignment of a real or complex number to each point in a way which varies smoothly as the point varies over the manifold. It can be shown that all relevant information about the manifold is contained in this family of functions. The essential point is that functions belonging to this family can be (1) added to each other, (2) multiplied with each other and (3) multiplied by real or complex numbers; and all these operations satisfy suitable axioms (analogous to those of the usual addition and multiplication of numbers). We say that the family of smooth functions on a manifold forms an *algebra*. Here the term 'algebra' does not denote a branch of mathematics, but rather a mathematical object studied in algebra understood as a branch of mathematics. It is another example of transmutations of mathematical concepts (in this case of the concept of algebra) in the course of the history of mathematics.

Both methods of doing geometry are equivalent, and both have their own advantages. The method of maps and atlases is easier for standard calculations, but the method of the algebra of functions gives us a deeper insight into the structure of manifolds, and is more elastic as far as further generalisations of the manifold concept are concerned. The latter property is extremely important. This is why a lot of modern work in differential geometry is done with the help of functional algebras rather than with the help of maps and atlases. When we pursue this line of research, we can forget about coordinates and construct the geometry of a manifold entirely in terms of the algebra of functions. Since functions on a manifold are multiplied in a commutative way (i.e., if f and g are functions on a manifold then $fg = gf$, in strict analogy with multiplication of real numbers, e.g., $3 \times 7 = 7 \times 3$), the geometry of standard manifolds can be called *commutative geometry*. Now, we can make use of the 'elasticity' of this method, and go one step further by claiming that any algebra, not necessarily a commutative one, defines a certain space. In this sense, we can speak about a *noncommutative space*. Although such a space is often difficult to visualise, the method itself turns out to be very fruitful both in pure mathematics and in its applications. Noncommutative geometry is a vast generalisation of standard geometry. Some 'sets' or 'objects', that do not surrender to usual geometric methods, are perfectly workable as noncommutative spaces.

One of the most striking peculiarities of noncommutative spaces is their totally global character. In principle, no local concept can be given any meaning in the context of noncommutative geometry. Typical examples of local concepts are those of 'point' and neighbourhood of a point. In general, noncommutative spaces do not consist of points. Let us look more closely at this peculiarity of noncommutative geometry. The basic idea of standard (commutative) geometry, in its map and atlas approach, is that every point is identifiable with the help of its coordinates. When we prefer the functional method, we can identify a point with all functions that vanish at this point. All such functions form what in algebra is called *maximal ideal* of a given algebra of functions. In other words, points are identified with maximal ideals of a given functional algebra. And now a surprise! It can be shown that for strongly noncommutative algebras there are typically no maximal ideals. In such circumstances, there is nothing that could be identified with a point. Similar reasoning could be used with respect to other geometrical local concepts (like neighbourhoods).

Nonlocality of noncommutative spaces has paramount consequences not only for geometry itself, but also for foundations of mathematics in general. It is often claimed that set theory can serve as a logical basis for all other mathematical theories. Of course, the crucial concept in set theory is that of 'belonging to a set'. For this concept to have a meaning, one must be able to identify elements of the collection under consideration by means of a denumerable family of properties. In noncommutative spaces, in general, there is no such possibility; one cannot distinguish elements of such a collection from each other by means of a denumerable family of properties. To grasp the meaning of this rather technical expression, let us imagine that one meets two twin sisters who have all properties the same, including their location in space and in time. One would certainly consider them as a single person. In an analogous way, within the noncommutative context no two 'individuals' are distinguishable from each other. Having formulated this unexpected result, Alain Connes concludes: 'The noncommutative sets are thus characterised by the effective indiscernibility of their elements.'[†]

It goes without saying that if we apply this 'strange' mathematical theory to model physical reality, we obtain a world with quite

[†] See p. 74 of Connes (1994).

unexpected properties. Perhaps we can now better understand the previous remark that noncommutative geometry would preserve its philosophical significance even in the case that it turns out to be inadequate (perhaps not exotic enough) to model the fundamental level of physics. A totally nonlocal world with no individual entities will certainly fascinate philosophers, even as a mere possibility, by opening before them an entirely new field for conceptual analysis. But perhaps this is also what we need in physics. So far our well-behaved mathematical structures have failed to implement our most cherished dream – the unification of gravity and quanta. It is worthwhile to try with a new toolkit of mathematical structures in hand.

5.6 PREGEOMETRY AND THE ORIGIN OF SPACETIME

The most obvious consequence of employing noncommutative geometry to model the fundamental level of physics is that on this level there could be no usual concepts of space and time (similar conclusions follow from some other approaches to quantum gravity). Indeed, space is but a collection of points, and time is but a collection of time instances, and both these concepts, as strictly local, have no meaning in the conceptual setting of noncommutative geometry. How can we imagine geometry without space or spacetime? The fact is that noncommutative geometry uses many methods typical for geometric reasoning (this is why it is called *geometry*), and it reproduces standard geometry when a suitable 'transition to a limit' is performed. This is the reason why it is sometimes called an example of *pregeometry*. I propose that we stick to this name – it well reflects both the novelty of the concept and our ignorance of its deeper nature. And the origin of this name introduces us well to the heart of the problem.

The dispute concerning the origin of spacetime has a long history in relativistic physics. Einstein himself spotted the problem as soon as he wrote down his famous equations of gravitational field. They present gravity field as a deformation of spacetime geometry caused by the matter distribution in it (this is often expressed by saying that 'gravity is a curvature of spacetime'): on the left-hand side of

Einstein's equations we have an expression describing the spacetime geometry, and on the right-hand side of them an expression describing the matter distribution. Although the two sides are connected by an equality sign, the question arises: what is 'physically prior' spacetime geometry or matter distribution? Motivated by the whole of the nineteenth-century philosophical tradition, Einstein himself attributed priority to material bodies. In his opinion, it is the matter distribution that generates spacetime.

Thinking about that, Einstein tried to make his 'principle' more precise. Matter distribution in spacetime is something global, and his field equations describe spacetime geometry in a purely local way (field equations from their very mathematical nature are local equations). Therefore, global properties of spacetime should generate, or entirely determine, its local properties. Einstein called this postulate *Mach's principle*. In fact Einstein's laborious road to his general theory of relativity was strongly motivated by the set of ideas developed by the influential nineteenth-century physicist and philosopher, Ernest Mach. In his famous 'bucket experiment' (which was only a thought-experiment) Mach argued that local inertial properties inside the rotating bucket would change if its walls are made 'several leagues thick'.

The problem goes far beyond a 'relativistic technicality'. Modern physics started its triumphant progress as soon as Galileo succeeded in isolating the free fall of a stone from the entanglement of interactions forming the structure of the Universe. How was it possible? Is the whole of the Universe the sum of its parts such that one of them (the free fall of a stone) can be isolated from the rest without substantially damaging the whole? There are strong reasons to believe that the essence of the physical method is inseparably connected with the question of interplay between local and global aspects of the Universe.

When Einstein finally created the general theory of relativity, he was convinced that his field equations contain Mach's principle in their very mathematical structure. It came to him as a surprise when, in 1917, the Dutch astronomer, Wilhelm de Sitter, discovered a new solution to field equations. The new solution described a universe with perfectly determined structure of spacetime but with vanishing matter density. In de Sitter's universe there is no matter but there is a well-determined spacetime structure. This evidently anti-Machian property initiated long-lasting disputes about the place of Mach's principle in General Relativity.

On the other hand, there was an effect in General Relativity that seemed to support Machian ideology. In 1918 Josef Lense and Hans Thirring calculated what can be regarded as an implementation of Mach's interpretation of the 'bucket experiment' into the mathematical structure of General Relativity. They demonstrated that the agglomeration of matter around a certain locality clearly changes the behaviour of test bodies in this locality. It is interesting to notice that recent measurements performed using NASA's LAGEOS satellites beautifully confirmed the predictions made by Lense and Thirring.

As discussions around these matters went on, it became more and more clear that Mach's principle is only partially implemented in General Relativity, in the sense that the global distribution of matter influences the local properties of spacetime, but it does not determine them fully. There is a simple argument demonstrating that no physical theory based on ordinary geometry can fully implement Mach's principle. Every curved space can always be locally approximated by flat space. We experience this geometric law in our everyday life; it took quite a lot of time and effort to convince people that their home (the planet Earth) is not flat space but a spherical globe. A little bit more precisely: every differential manifold can be locally (in a neighbourhood of a given point) approximated by the flat tangent space to the manifold (at this point). Whatever we do to change the global structure of spacetime, it will always be 'locally flat'. We can modify our local physics as we want, but this 'absolute element' will always be hidden in it as long as we employ ordinary geometry to shape properties of spacetime.

If there are serious difficulties in creating 'spacetime out of matter', why not try the reverse, namely to create 'matter out of spacetime'? Such an idea was proposed by John Archibald Wheeler in the sixties of the last century. The idea was to generate all the spectrum of elementary particles, known at that time, from various geometric configurations of spacetime. Wheeler baptised this idea with the name *geometrodynamics*. Spacetime is not only a geometric stage on which physical processes develop. It has truly dynamical character and is the foundation of everything that exists. Wheeler and his coworkers put a lot of effort and ingenuity to implement this programme, but the results – besides elaborating new mathematical tools – were rather modest. When Wheeler finally realised that the programme would not work, he modified his idea. It is not geometry – he claimed – that generates matter, but a more primordial stuff for which he coined the name 'pregeometry'.

What is this mysterious pregeometry? The concept was vague, as Wheeler expressed this himself in his famous textbook on relativity by saying that pregeometry is 'a combination of hope and need, of philosophy and physics and mathematics and logic'.[†] Wheeler played with several possibilities. Among others, he considered the idea of elementary bits of information as fundamental building blocks of physical reality; he even coined the saying: 'every *it* from a *bit*'. He also explored the idea of constructing pregeometry out of elementary logic (logic of propositions). He argued that logic is the only branch of mathematics that can 'think about itself'. A common denominator of all these ideas was that everything had to be made out of pregeometry whatever its nature, of which the usual spacetime geometry is but an approximation.

All these conceptions appear seldom now in the centre of scientific papers, but the ideas underlying them are still alive. Many theoreticians believe that on the Planck level the ordinary spacetime dissolves into something more fundamental, point-like events disappear

[†] The textbook is Misner, Thorne and Wheeler (1973), see p. 1203.

and the usual interplay between 'local' and 'global' is replaced by a new structural pattern. Having this in mind, it is hard not to think about noncommutative spaces as candidates for pregeometry. Let us explore this possibility.

The most demanding formulation of Mach's principle postulates that all local properties of spacetime should be *fully* determined by the global structure of spacetime, and we have seen that the rigid structure of usual geometric spaces, which have to be locally flat, prevents this from happening. Indeed, any local neighbourhood of any point in such a space is, at least to some extent, determined by the fact that it is approximately flat, irrespectively of what happens at large distances. In the noncommutative environment this situation changes radically. Local properties simply do not exist. They are totally engulfed by the global structure which is all that exists. In this sense, a physical model, based on noncommutative geometry, is indeed fully Machian; in such a model the whole of physics is entirely determined by the global world structure. All localities, together with spacetime and all anti-Machian effects, emerge only when noncommutative pregeometry undergoes a transition to the familiar world of commutative geometry.

If we agree that noncommutative geometry, as a mathematical stuff of Planckian physics, has many properties that would qualify it – in Wheeler's eyes – as a perfect candidate for pregeometry then we are entitled to say that two so far rival philosophies – Mach's principle and Wheeler's geometrodynamics – are nicely unified on the fundamental level. 'Noncommutative pregeometry' has both Machian and Wheeler-like properties. These two philosophies become rival only in our world of broken primordial symmetries.

5.7 MYSTERY OF IDENTITY

One of the central concepts of Western philosophy is that of individual. In what sense can we meaningfully speak of an individual in the noncommutative context where neither space nor time localisation obtain? This is an especially troubling question if we remember that there are many philosophers who claim that it is the localisation in space and time that constitutes the identity of objects. It was Gottfried Wilhelm Leibniz who formulated his famous Principle of the Identity of Indiscernibles: two entities that are identical in every respect (localisation in space and time included) are the same entity.

The concept of individuality preserves its importance in classical physics, but already in quantum field theories serious difficulties connected with this concept appear. In our macroscopic world there is a sense 'in which we can talk about this object (object 1, say), whatever properties it may have, as opposed to that object (object 2, say), whatever properties this second object may have'.[†] Medieval philosophers called this 'haecceitas', which might be crudely translated as 'thisness'. Every macroscopic object has its 'thisness', but elementary particles, belonging to the fermion family (like proton or electron which are constituents of ordinary matter), do not have this property. They are identical in the strong sense of this term: if we exchange two of them with each other, literally nothing changes in the system. No 'thisness' can be ascribed to them. In this sense Hermann Weyl used to say that we cannot ask alibis of electrons. Should we guess that this effect is but a remnant of the totally nonlocal character of the fundamental level? If so, we would have a striking gradation: on the fundamental level there is no individuality at all; in quantum physics, to use Whitehead's expression, electrons exist 'like a melody' rather than like small pieces of matter; in the macroscopic world of classical physics there exist pieces of inert matter immersed in space and time, but only with the appearance of life does the intensity of the individualisation process became more pronounced, to reach, in human beings, an unprecedented degree of sophistication. We can truly say that only in human self-consciousness has cosmic evolution acquired this unique property of being able to ask questions addressed to herself.

[†] See pp. 16–17 of Teller (1995).

5.8 NONLOCAL PHENOMENA

In the preceding section we have suggested that problems with the identity of particles in contemporary quantum field theory might be a remnant of the nonlocal Planckian era. Perhaps can we find some other such remnants in our present physical theories? We should evidently look for them in those domains in which there are some troubles with localisation. And indeed there are known in physics certain phenomena of typically nonlocal character which either remain unexplained or require some additional (often artificial) hypotheses to explain them. Let us mention two of them.

First is the so-called Einstein–Podolsky–Rosen (EPR) paradox. In 1935 these three authors proposed a gedanken experiment aimed at showing that quantum mechanics is an incomplete theory. Suppose that an atom, in a laboratory in London, emits two electrons which travel far away from each other, for instance one to Tokyo, another to New York. Electrons have a certain property called by physicists spin. According to the laws of quantum mechanics spin can assume only two values; let us call them: spin 'up' and spin 'down'. You could imagine that an electron is spinning (rotating) around its axis, and if it spins to the left, its spin is 'up', and if it spins to the right, its spin is 'down'. In fact, the concept of rotation in space is inadequate with respect to electrons, but this naïve picture could help you to grasp the idea. Quantum mechanics teaches us that if two electrons once interacted with each other (e.g., were emitted by the same atom), they cannot have identical spins: if one electron has spin 'up', the other one must have spin 'down'. Moreover, electrons have spin in a very peculiar sense, quite unlike when you have ten pounds in your pocket. Ten pounds are in your pocket irrespectively of whether you touch them, or not. An electron, when left to itself, has no determinate spin; only when spin is measured, it is registered as either 'up', or 'down'. Suppose now that in a lab in Tokyo the spin of one of our electrons is measured, and the result is 'down'. If then, immediately after that, physicists in New York would measure the spin of the other electron, their result must be 'up'. Instead of two electrons in Tokyo and New York you can imagine two electrons at two opposite sides of a cluster of galaxies; the reasoning will be the same. How could the electron know the result of the spin measurement of its twin brother on the other side of the cluster of galaxies immediately after the measurement has been done? It looks as if spatial distances do not exist for electrons. Einstein and his associates claimed that such a nonsensical result demonstrates the fact that quantum mechanics is not a complete theory.

What in Einstein's time could be only a thought-experiment, nowadays, with our present technology, has been many times performed. And it has turned out that the 'crazy' prediction of quantum mechanics is correct. 'Quantum nonlocality', as it is now called, is an empirical fact, and physicists have to live with it. Let us emphasise that 'quantum nonlocalities' (the EPR experiment is one of them) are impeccably deduced (with strict mathematical rigour) from the principles of quantum mechanics. In this sense, they are not paradoxical. It is only our intuition that cannot easily accept that two elementary particles can interact with each other with no mediation of space and time distances separating them.

Another typically nonlocal phenomenon appears in cosmology, and is known as the horizon problem. As early as in 1948 George Gamow calculated that the early Universe was filled with very hot electromagnetic radiation, and predicted that a remnant of this radiation should now be present in the form of radiation very uniformly spread in space at the temperature of the order of a few degrees Kelvin. This radiation field was discovered in 1965 by Arno Penzias and Robert Wilson (who were honoured with the Nobel Prize for this discovery), and later on investigated, in a very detailed way, by two satellites: COBE and WMAP (the latter one is still in orbit doing its work). Owing to this discovery cosmologists have acquired insight into the physical processes driving cosmic evolution soon after the Big Bang. But here I am interested in another aspect of this discovery. Recent measurements of the microwave background radiation (this is its commonly accepted name) have disclosed that its present temperature of 2.725 degrees Kelvin is almost exactly the same throughout the entire sky. Deviations from the average temperature are only about as 1 to 100 000. And now another piece of seemingly innocent information. If we look at two places in the distant sky, separated from each other by an arc of 30 degrees, say, then from the very beginning of the Universe there was not enough time for a physical signal to go from one of these places to another. These two places in the sky are, to use cosmologists' parlance, causally totally disconnected: no causal interaction can connect them. If so, how could the microwave background radiation 'synchronise' its temperature at these two places to the same value with such an incredible precision? How can parts of the Universe that have never been in causal contact with one another be characterised by exactly the same value of the microwave background radiation temperature? This question is usually answered by the so-called inflationary model of the very early Universe which presupposes that in a tiny fraction of the first second after the Big Bang, a very rapid expansion was superimposed on the 'ordinary' expansion typical for the standard model of the Universe. Owing to this 'inflationary phase', the entire Universe, visible at present to our telescopes and radio telescopes, was once squeezed into a very small volume, all parts of which could be in causal contact. But the inflationary model is an extra cosmological hypothesis that has never been observationally confirmed in an independent way.

All these phenomena find their natural explanation in nonlocal versions of quantum gravity. If the fundamental level is totally nonlocal, it is no wonder that some quantum phenomena – such as the EPR type of experiments – which are rooted in that level, exhibit nonlocal behaviour. They are but the tip of the iceberg of the 'fundamental nonlocality' which somehow survived the 'phase transition' to usual physics. The paradoxical behaviour of two electrons from the EPR experiment is due to the fact that these two electrons, by their very quantum nature, explore the fundamental level in which there is no concept of distance in space. They reveal their presence in space only when they are forced to interact with a measuring device belonging to our macroscopic world and subject to all its spatial constraints.

To explain the horizon problem, we should adopt another perspective and regard the 'fundamental nonlocality' as situated 'in the beginning', in the Planck era, at close vicinity of the Big Bang. Two photons of the microwave background radiation, now far away from each other, by their very quantum nature, are still exploring the atemporal Planck era in which the concept of spatial distance is meaningless and in which everything is in contact with everything. It is only when these photons are forced to interact with our registering radio-antennas that they disclose their space separation, meaningful only in our macroscopic world.

If we agree that our macroscopic world is but a shadow of the atemporal and aspatial fundamental level, we should expect in it some traces of its very nonlocal roots. And the fact that such traces indeed do exist should be regarded as an argument on behalf of our hypothesis.

5.9 TIME AND DYNAMICS

Let us further explore a nonlocal world. Our intuition tells us that, since motion is but a change of localisation in time, there can be no motion without time and localisation. Consequently, on the fundamental level, governed by noncommutative geometry, there can be no dynamics. It must be a static world. However, by now we should have learned that it is our intuition that should be shaped by mathematical reasoning, and not vice versa. Without the usual concepts of space and time there can be no usual dynamics, but this does not mean that, without these concepts, there can be no dynamics in a generalised sense.

In standard physics to describe motion we often use vectors. For instance, the velocity of a moving particle is a vector. Since vectors are local beings (a velocity vector is always attached to a point in space), they do not appear in the noncommutative setting. But even in standard physics we sometimes describe motion (or dynamics, in general) by using vector fields (e.g. in the theory of dynamical systems).

Such a vector field is a global entity, in spite of the fact that it consists of individual vectors. It describes a state of motion in its totality. And it is this global aspect of vector field that can be generalised to the noncommutative context. Such a noncommutative counterpart of vector field is called a *derivation*. We can think of it as a vector field out of which everything connected with the individuality of vectors has been erased. And exactly derivations can be used to model generalised dynamics with no usual concepts of space and time.

Indeed, 'there are more things in heaven and earth than are dreamt of in your philosophy'. I think, however, that we are underestimating philosophers. The concept of a 'timeless dynamics' is known, for a long time, in the Christian philosophy of God. It was even a common teaching among medieval masters. The reasoning was simple. God has all perfections in the maximally possible degree. Therefore, nothing can be either added to, or subtracted from, God's perfection. Thus God has to be immutable. But this does not mean that He remains inactive (in the Middle Ages God was certainly 'He' and not 'She or He'). On the contrary, He is full of activity, or the activity itself. How is this possible? This well fits into another traditional philosophical (or theological) doctrine, having its roots in Ancient Greece, and later on elaborated by St Augustine and Boethius (a Latin thinker from the turn of the fifth and sixth century AD). God does not exist in time, but in eternity, and eternity should not be understood as time without beginning and end, but rather as existence outside time. And such an existence is not static. As Boethius said in his *Consolation of Philosophy*, 'eternity is the entire and perfect possession of endless life at a single instant'. As compared with this 'maximally concentrated' dynamism, our own activity, diluted in time, might seem weak and painful.

It is only with the advent of Newtonian physics and its enormous successes that the idea of eternity, as time flowing from 'minus infinity' to 'plus infinity', started to conquer our intuition and imagination. Newton's idea of absolute time which 'flows uniformly with no regard to anything external' led him to the view that absolute time is a 'sensory' (an organ) of God's eternity; likewise absolute space is a 'sensory' of God's omnipresence. That is to say that even God exists in absolute time and absolute space. These ideas are so heavily laid down on our intuition that we react to the traditional idea of eternity as existence outside time (and space) with unwillingness and resistance. This reaction is even strengthened by our modern views on the history of the Universe and life. It is the theory of evolution that has added a strong temporal dimension to our image of the world. Some modern philosophers, like for instance Alfred North Whitehead, and theologians, like Paul Tillich, were so fascinated with the idea of flowing time that they opposed the traditional doctrine on Boethian eternity, and preferred to see God – especially Whitehead – as participating in the adventure of temporal processes. This tendency is also present in current Christian theology (see the essay by Polkinghorne in the present book); however, this does not change the fact that many approaches to quantum gravity suggest something different: on the fundamental level time either does not exist, or has drastically different properties from what we are accustomed to in the macroscopic world. Anyway, the predominant feature of this world is its transience determined by the unidirectional flow of time. And as Gerald Whitrow once cleverly remarked, 'Any theory which endeavours to account for time completely ought to explain why it is that everything does not happen at once'.[†] In the noncommutative perspective timelessness seems natural, and the fact that our world is 'temporally extended' requires justification.

5.10 TOP-DOWN AND BOTTOM-UP

In the rest of this essay, I will go back to the traditional Christian doctrine of creation, and try to enrich it with insights coming from the conceptual possibilities opened by our search for a theory of quantum gravity. There are three such concepts, closely related to the idea of creation, that find a new elucidation within the context of quantum

[†] Whitrow (1975).

gravity, namely the concepts of causality, chance and finality (design). As we have seen, this context is powerfully shaped by its two properties: timelessness and nonlocality. Therefore, if we want to obtain suitable counterparts of these three concepts, we must, first of all, strip them of their usual involvement in time and locality. Let us begin with causality.

Creation can be understood as *sui generis* a causal nexus between God and the world. In this context, God is often said to be the Prime Cause of the world. It is obvious that the very idea of a chain of secondary causes subordinated to the Prime Cause was shaped by analogy with everyday experience in which we often meet several linearly successive causes producing a single effect. Precisely at this point we have to revise our theological ideas. In the light of what has been said in the preceding sections, we should seriously take into account the possibility that the concept of causality does not necessarily presuppose a local interaction between the cause and its effect, and does not necessarily presuppose a temporal order.

The problem of causality is notoriously difficult even in the macroscopic world. I do not intend to immerse myself into endless philosophical discussions; I shall stick to its most commonsense aspects. It seems obvious that causation presupposes a separation in space and a temporal order. If A causes B then A and B are distinct; otherwise we would have a self-causation. If A causes B then A precedes B; otherwise, B could be the cause of A. However, the separation in space and a temporal order do not form a sufficient condition for causality. For two events to remain in a causal nexus, a certain dynamical interaction must exist between them. There are good reasons to think that if we liberate causality from its involvement in separability in space and ordering in time, it is exactly this dynamical interaction that remains, and this seems enough for justifiably speaking about causal dependence.

Let us come back to the EPR experiment. Both electrons, the one in Tokyo, and another in New York, find themselves beyond any physical influences spreading in space and time. But it seems that there is a sense in which one can say that the measurement of an electron's spin in Tokyo is a cause of the result obtained by the measurement of an electron's spin in New York. A certain dynamical nexus surely exists between the two in spite of the fact that no physical signal was transmitted from Tokyo to New York. It is this kind of 'generalised causality' that should be expected in the nonlocal world of Planck's era.

In recent discussions on causality, two kinds of causality have been distinguished. The causality, as it most readily functions in physical sciences, is a *bottom-up* causality. It occurs when a property of a whole results from interactions between its constituent parts. This kind of causality has a strong reductionist flavour: a property of a whole is reduced to its 'elementary building-blocks'. But great progress in investigating complex systems, in recent decades, has taught us to admit the existence of the reverse kind of causal influences. They occur when the state of a system as a whole determines the behaviour of its constituent parts. For instance, the whole can regulate or constrain the behaviour of its parts, or determine boundary conditions for their dynamics. This kind of causality is called *top-down* causality. It has a holistic flavour. In this perspective, parts are subordinated to the totality.

It is rather obvious that these two kinds of causality do not necessarily exclude each other. It would not be difficult to quote examples when they act in a complementary way, but nevertheless, from the conceptual point of view, they are different. This is so in our macroscopic world, and how do things look like, in this respect, in a nonlocal regime? At first glance it would seem that, since the nonlocal world is totally holistic, one can meaningfully speak in it only about a top-down causality. It cannot be denied that the nonlocal world is a holistic entity, up to the extent that the concept of a component part has hardly any meaning in it. Therefore, if it is a relationship between the whole and its parts that distinguishes top-down causality from bottom-up causality, then in the nonlocal stetting the very distinction between these two kinds of causality disappears.

It is interesting to notice that the issue of bottom-up causality and top-down causality has recently emerged in theological discussions on 'God's action in the world'. The problem was how can God act in the world without violating or suspending laws of nature (in a 'non-interventionist way' as theologians are used to say). It seems that God could choose either the bottom-up channel, or the top-down channel to act on the world. Both standpoints have been defended.

The bottom-up channel was defended by, among others, Robert Russell.[†] In his view, this channel is possible only 'if nature, according to science, is not an entirely closed causal system'. This is indeed the case, because at the very 'bottom' of physics, on the quantum level, macroscopic determinism is replaced by quantum indeterminism. This creates a place for God's action. In this approach God can act purposefully, through quantum indeterminacies, without disrupting or violating the laws of nature. 'God's special action results in specific, objective consequences in nature, consequences which would not have resulted without God's special action. Yet, because of the irreducibly statistical character of quantum physics, these results would be entirely consistent with the laws of science.'

Some other theologians, Arthur Peacocke among them, prefer the top-down channel.[‡] Peacocke's approach 'is based on the recognition that the omniscient God uniquely knows, across all frameworks of reference of time and space, everything that is possible to know about the state(s) of all-that-is, including the interconnectedness and interdependence of the world's entities, structures, and processes. By analogy with the operation of the whole-part influence in real systems, God could affect holistically the state of the world (the whole in this context). Thence, mediated by the whole-part influences of the world as-a-whole (as a System-of-systems) on its constituents, God could cause particular events and patterns of events to occur which express

[†] Russell (1998).
[‡] Peacocke (1999).

God's intentions. These latter would not otherwise have happened had God not so intended.'

It would be ridiculous even to try to construct a model of God's interaction with the world on the basis of our attempts to create a theory of quantum gravity (or any other mathematical model), but it would be equally unreasonable not to notice that the above theological speculations seem less extravagant if read with the consciousness that even more abstract ideas can be found in the most exact among exact sciences. Mathematicians and theologians have something in common: both are speaking of what evades our senses but what cannot be otherwise.

5.11 CHANCE AND PURPOSE

The theory of probability plays an important role in contemporary science. Quantum mechanics and quantum field theories, dealing with elementary particles and elementary forces, are essentially probabilistic theories, and there are strong reasons to believe that the looked-for fundamental physical theory will be probabilistic as well.

To determine the probability of a certain event, we must first ascribe a number to each possible event of the considered class. For instance, to every possible outcome of throwing dice, we ascribe the number 1/6. In other words, we define a function on the set of all possible outcomes of throwing dice; this function is called a *distribution function* (in this case, it is a constant function: to every result of throwing dice we ascribe the same number, namely 1/6). From our point of view, it is important that it is a function. The set of all possible events, called *elementary events*, on which the distribution function is defined, is called the *probability space*. In some cases, when we want to have an elegant mathematical structure, we can define a probability space entirely in terms of an algebra of functions. The point is that this algebra of functions must have certain properties that relate it to the concept of probability. Algebras with these properties are called *commutative von Neumann algebras*; they are commutative in the same manner as we have seen for functions on a manifold in Section 5.5. As we remember, differential manifolds can be defined either in terms of coordinates, or in terms of algebras of functions. Both methods are equivalent, but the latter leads to a generalisation; namely, to noncommutative spaces. In the present case, analogously, probability spaces can be defined either in terms of elementary events and probabilities ascribed to them, or, at least in some cases, in terms of commutative von Neumann algebras. And again the latter method is open for a generalisation. If we drop the commutativity assumption, we obtain what should be called *noncommutative probability space*. In fact, a well-known version of this approach to probability theory was developed by Dan Voiculescu, and is called by him *free probability theory*.

The usual concept of probability presupposes individual events – the elements of the probability space (e.g., a single event of throwing dice). How could the concept of probability live in an environment devoid of all local concepts? We had the same problem with noncommutative dynamics. How is it possible with no local concepts? To help our imagination we must consider together dynamics and probability. If a dynamical nexus is possible between nonlocal phenomena, it could be somehow 'biased' toward this rather than that state of the system. I have used the term 'state' on purpose since 'state' has a certain global connotation. The entire system can be in this or in that state. For instance, the entire audience in a spectacle can be in a state of a sleepy dullness, or in a state of excitement. In noncommutative geometry, states can, to some extent, be substitutes for points. And it is interesting to notice that in noncommutative geometry the same kind of states play an important role in both noncommutative dynamics and noncommutative probability. It looks as if every dynamics (in a generalised sense) is probabilistic (in a generalised sense), and *vice versa*, every probability has a dynamic aspect. Unfortunately, we are unable to go into technical details of this interesting domain, but some philosophical aspects of noncommutative probability certainly deserve to be briefly discussed.

The concepts of chance and purpose always had great significance for philosophy and natural theology. We usually think about these concepts in terms of probabilities. If an event of a very small probability happens, we are inclined to say that it has either happened by chance or has been chosen on purpose by an intelligent agent.

In classical natural theology, the low probability of an event was often considered as a 'gap' in the ordinary course of nature that had to be filled in with the direct action of God. Many of the so-called proofs of God's existence were based on this strategy. An echo of this type of argument can be heard today in discussions around the Anthropic Principles. As is well known, the initial conditions for the Universe had to be extremely 'fine tuned' in order to produce a world friendly for the evolution of life. Roger Penrose estimates that in order to produce a universe resembling the one we inhabit the Creator would have to aim for an absolutely tiny fraction of all possible universes – the fraction of an order of one over ten to the power ten to the power one hundred and twenty-three. Penrose is using here the image of the Creator as a metaphor to evoke in the reader an astonishment of this minute probability, but some apologists treat this as a real argument on behalf of God's existence. On the other hand, some antagonists, wanting to neutralise this type of argument, assume that there exists an ensemble of all possible universes with all possible combinations of initial conditions (the so-called multiverse). Then – they claim – no wonder that we are living in the very special Universe, because only within such a Universe is life possible. For us the probability to live in this very special Universe is nearly one. God is not indispensable to make any fine tuning of the initial conditions. In this way, high probability becomes the principal competitor of God.

All arguments of this type tacitly assume that the classical concept of probability is valid on all levels of reality and, moreover, that God knows only classical probability calculus. However, there are reasons to believe that the classical concept of probability, just as many other concepts, has only a restricted domain of applicability, outside of which we should be open to various possibilities. The existence of the noncommutative generalisation of the standard concept of probability should be treated as a serious warning in this respect. I am far from claiming that we should use the noncommutative generalisation of probability in our speculations about anthropic principles and the multiverse. I only insist that we should not absolutise even our most 'obvious' concepts, the concept of probability included.

By the way, if God, as an Infinite Being, is not interested in anything less than infinity, then the multiverse need not be God's rival, but rather one of His best distractions.

5.12 PURPOSE AND EVOLUTION

A dispute around so-called intelligent design is, in recent months, a hot topic in the media. Intelligent design is an anti-evolution belief claiming that no purely naturalistic explanation of the phenomenon of life is possible, and that this phenomenon can only be explained by intelligent causes. Religious advocates of this standpoint argue that the theory of evolution substitutes chance for the Creator which – they claim – is inadmissible from the theological point of view. To immerse ourselves into details of this discussion would blow the limits of the present essay, but I cannot resist making a few comments related to our main theme.

Our concept of design is heavily laden with the idea of temporality: pursuing a purpose or a goal means to choose a goal in the future, to determine suitable means to reach it, and to initiate a chain of actions in this direction. The full knowledge of the outcome of this process is possible only if it is strictly deterministic, and if its unfolding does not too sensitively depend on the initial conditions. In this conceptual setting there is indeed the contradiction between purposeful action and purely random events or chance. But it is also a strongly anthropomorphic description of the situation. Things look differently if looked upon from the atemporal perspective. Moreover, God's knowledge of outcomes of human decisions need not ruin our freedom. God contemplates these outcomes 'by inspection' from God's atemporal perspective, in which everything has been already completed.

As transparently expressed by Ernan McMullin, 'Terms like "plan" and "purpose" obviously shift meaning when the element of time is absent. For God to plan is for the outcome to occur. There is no interval between decision and completion. Thus the character of the process which, from our perspective, separates initiation and accomplishment is of no relevance to whether or not a plan or purpose on the part of Creator is involved'. Knowledge of the material world by the atemporal God 'is not discursive': God does not need to infer or to compute future states of the Universe from the knowledge of its earlier states. He knows what for us is the past and the future by inspection, and in his planning there is no element of expectation.[†] In this perspective, what we call purely random events can be inscribed into God's planning the world.

This shift of meanings of such concepts as probability, chance and purpose has obvious consequences as far as the intelligent design dispute is concerned. McMullin writes: 'It makes no difference, therefore, whether the appearance of *Homo sapiens* is the inevitable result of a steady process of complexification stretching over billions of years, or whether on the contrary it comes about through a series of coincidences that would have made it entirely unpredictable from the (causal) human standpoint. Either way, the outcome is of God's making, and from the Biblical standpoint may appear as part of God's plan.'[‡]

McMullin is right when he says that the Christian doctrine can, in principle, be interpreted in either way: as a series of direct interventions in the course of natural events, or as a natural evolutionary process. But the correct theology is obliged to take into account what science has to say to us in this respect. And the verdict of science in this respect is clear. The Universe we live in is an evolutionary process, and the thread leading from the plasma of primordial stuff, through chemical elements, galaxies, stars and planets, to more and

[†] McMullin (1997).
[‡] McMullin (1997).

more complex systems, intelligent life included, is but a fibre in this overwhelming process. Any theology that would choose to ignore this magnificent process is a blind way to nowhere.

5.13 SCIENCE, THEOLOGY AND CONTEMPLATION

In the present essay, I often jumped from mathematical and physical models to theological speculations. In doing so I have certainly scandalised many scientists, who usually prefer keeping science and theology in two compartments strictly isolated from each other, and many theological purists who claim that any mixing of theology and science must lead, by not respecting their very different methodologies, to confusion and misunderstanding. I feel obliged to make a few clarifications.

Ian Barbour, in his often cited paper 'Ways of relating science and theology' classified all standpoints in this respect into four groups: conflict, independence, dialogue and integration.[†] Adherents of the conflict approach 'claim that science and theology make rival literal statements about the same domain, the history of nature, so that one must choose between them' (interestingly, Barbour includes into this group both scientific materialism and biblical literalism). Those who proclaim independence argue that science and theology have created 'two jurisdictions and each party must keep off the other's turf'. Adherents of the dialogue strategy usually begin with methodological distinctions that resemble 'independence', but they end with a 'concern of compatibility, consonance and coherence'. 'Integrationists' either claim that 'scientific theories may contribute to reformulation of theological doctrines', or hope that both science and religion 'may contribute jointly to the formulation of a systematic synthesis', a sort of coherent world-view composed of scientific and theological elements.

From the analyses carried out in the present essay it is clear, I hope, to which of the above four groups of opinions I would strongly object. To which of them would I adhere? In general, to none of them.

[†] Barbour (1988).

In such complex matters, to succumb to the instinct of classifications could easily lead to dangerous simplifications. Each question should be pondered individually, having only in mind some general principles (which, by the way, should be elaborated by working with particular problems rather than assumed *a priori*). This is why below I will briefly reflect on the method I have employed in the present essay when balancing on the interface between some conceptual problems met in our search of quantum gravity, on the one hand, and some philosophical and theological speculations, on the other.

I would describe my strategy of dealing with this delicate interface as an *intellectual contemplation of conceptual schemes*. The word 'contemplation' (*contemplatio*) is the Latin translation of the Greek 'theory' (*theoria*), and I am using it here in this original sense. Contemplation was an important method in the philosophy of Plato. In his view, the soul, by means of contemplation, may ascend to knowledge of divine Forms or Ideas. It is especially fitting if we remember that, in Plato's teaching, mathematical objects or structures belong to the inhabitants of the World of Ideas. The Christian understanding of contemplation, as a part of mystical experience, is a later derivative of the original Greek concept, and as such it is not intended here.

My type of contemplation I qualify as an *intellectual* contemplation. This is to underlay its rational character. I am certainly not hunting for any 'deeper insight' by means of intuition and feelings. I am dealing with mathematical structures in a responsible manner; I am using logical tools in analysing the interaction of concepts, and the ways they are generalised. When I am on the verge of losing this secure track, I warn the reader openly. My leading idea is that the limits of rationality do not necessarily coincide with the limits of scientific method. The method of science explores the easiest kind of rationality, but I think that to give up, when we are confronted with more complex forms of rationality, would be an irrational move.

And I am contemplating in this way conceptual schemes rather than individual concepts. Concepts always form subtle networks;

they interact with each other in manifold ways; minor shifts of meanings at the one end of the network can result in a powerful transmutation of the whole. All these processes often participate in solving various problems, or at least posing new questions and illuminating old answers.

This sort of contemplation gives to the one who undertakes it a strong experience of participating in something that surpasses a single researcher. The sum of the efforts of many scientists, philosophers and theologians scattered throughout the world, now and in history, moves towards Something that reveals its many faces only if compelled to do so by cunning methods of logic and inventiveness. This Something has many names, but most prosaically is called Reality. The contemplating researcher belongs to it and participates in its strangeness.

In this essay, I have contemplated a scheme composed of the following pairs of concepts: local – global, temporal – atemporal, causal – dynamically dependent, random – purposeful, that emerge out of our search for the theory of quantum gravity. This conceptual scheme has been used to gain an insight into the traditional Christian doctrine of creation. I have strongly insisted that gaps in our present knowledge (and the lack of the commonly accepted quantum-gravity theory is certainly such a gap) cannot be used to argue on behalf of God's creative action. What I have done is just the opposite: from our hypothetical models that we construct to fill in the quantum-gravity gap, I have tried to learn a lesson that could help us to better understand the doctrine of creation. If creation is a continuous dependence of the world in its existence on God (a continuous 'giving existence' by God to the world) then there are reasons to believe that the roots of this dependence should be looked for on the most fundamental level of the world's structure. Consequently, the concepts born in our attempt to penetrate this level should be more useful to gain insights into the idea of creation than those elaborated in our interaction with the macroscopic world. However, we should be aware that even our most sophisticated concepts have only a limited field of application and are inadequate to grasp the Transcendental One.

In the present essay I often alluded to God and drew parallels between theological speculations and conceptual horizons opened up by new mathematical and physical theories. I am far from claiming that new developments in the evolution of conceptual schemes are able to catch the idea of God. We always should remember that all our concepts, even maximally abstract and generalised, refer to Demiurge, the product of our discourse and imagination, rather than to a truly transcendent God. In this sense, it would be reasonable everywhere in this essay to replace the word 'God' by the word 'Demiurge'. I do not do so only to conform to the commonly accepted way of speaking, and not to make my text too pedantic. Neither was I using a philosophy of God exclusively as a didactic tool, helping our imagination to grasp subtleties of quantum-gravity level. I wanted to achieve a certain critical distance with respect to our commonsensical intuitions and time-honoured mental patterns. This is desirable in both science and theology.

Good advice in this respect was given by Nicholas Saunders: 'What we really need in our research into divine action is to try out our hesitant models in all their details whilst recognising the inherent limitations of our human perspective.'[†]

BIBLIOGRAPHY

Barbour, I. (1988) Ways of relating science and theology, in *Physics, Philosophy and Theology: A Common Quest for Understanding*, ed. R. J. Russell, W. R. Stoeger, G. Coyne, Vatican Observatory Publications, pp. 21–48.

Callender, C. and Huggett, N. (2001), eds., *Physics Meets Philosophy at the Planck Scale*, Cambridge University Press.

Connes, A. (1994) *Noncommutative Geometry*, Academic Press.

McMullin, E. (1997) Evolutionary Contingency and Cosmic Purpose, *Studies in Science and Theology*, **5**, 91–112.

Misner, C., Thorne, K. and Wheeler, J. A. (1973) *Gravitation*, Freeman.

[†] Saunders (1999).

Peacocke, A. (1999) The sound of sheer silence: how does God communicate with humanity?, in *Neuroscience and the Person*, ed. R. J. Russell, N. Murphy, T. C. Meyering, M. Arbib, Vatican Observatory Publications/CTNS, pp. 215–47.

Russell, J. R. (1998) Special providence and genetic mutation: a new defence of theistic evolution, in *Evolutionary and Molecular Biology*, ed. J. R. Russell, W. R. Stoeger, J. Ayala, Vatican Observatory Publications/CTNS, pp. 191–223.

Saunders, N. (1999) Special divine action and quantum mechanics, *Studies in Science and Theology*, **7**, p. 190.

Teller, P. (1995) *An Interpretative Introduction to Quantum Field Theory*, Princeton University Press.

Whitrow, G. (1975) *The Nature of Time*, Penguin Books.

6 The nature of time

John Polkinghorne

Not many of us are perplexed about space. We can move around in it and its nature seems experientially obvious. Yet even in the case of spatial properties, the philosophically minded can deem the existence of reliable measuring rods, capable of metricating space, as not being as straightforward a matter as commonsense might suppose. Moreover, when physical cosmologists theorise about the Universe, they find that its vast spatial domains exhibit an intrinsic curvature, corresponding to General Relativity's account of the nature of gravity. There are certainly subtleties about the nature of space, which go beyond the expectations of everyday thought, but they are nothing like as perplexing as those we encounter when we attempt to think about the nature of time.

Time travel is not available to us and we have to take our experience of time 'as it comes to us', in the succession of those fleeting present moments which as soon as we experience them recede immediately into the inaccessible fixity of the past. Famously, St Augustine, meditating on temporality in the Confessions, said that as long as he did not think about it, he knew what time was, but as soon as he reflected on the nature of temporal flux, he began to be perplexed. To commonsense, the one thing that does seem clear is that time flows. Yet one of the central issues in the modern discussion of temporality is whether this is indeed the case, or whether our human sense of the flow of time is merely a trick of psychological perspective, and the fundamental reality of time is quite different. The resulting debate takes place between the proponents of what is called the block

On Space and Time, ed. Shahn Majid. Published by Cambridge University Press.

universe and those who support the picture of a universe of unfolding becoming.[†]

The former believe that the true reality is the totality of the spacetime continuum, the whole of cosmic history taken altogether. Classical relativity theory is formulated in terms of this 4-dimensional entity and Albert Einstein was a convinced block universe man, even writing to console the widow of a friend with the thought that her husband was 'still there' in that part of the spacetime continuum that contained his life span. Two principal arguments appealing to science have been advanced in support of this counterintuitive concept.

The first appeals to the undoubted fact that according to Special Relativity, two different observers moving in different fashions will make differing judgements about the simultaneity or otherwise of distant events. If two events, A and B, might be such that one observer judges A and B to be simultaneous, another observer judges A to occur before B, while a third observer judges B to precede A, does this not show that the distinction between past, present and future is purely subjective? I do not think so. In fact, a careful analysis of what is going on can show that relativity theory does not imply the necessary truth of the claim, implied by the block universe view, that the future already exists. In Special Relativity, each observer is certainly organising his description of past events in a different way, but for each of them the events so described are unambiguously past, since no observer can have knowledge of a distant event until it is in his past lightcone. That domain is invariantly defined, that is to say, all observers will agree about it. Thus the variations in the way that past history is described cannot be appealed to establish the already existing reality of the future.

The second argument is simpler in character. It simply notes that the equations of physics do not contain any representation of 'the present moment'. One might say that $t = 0$ has no special significance.

[†] Isham and Polkinghorne (1993).

That is why the classical relativists could work with a picture embracing the whole of spacetime. Yet, fundamental physics is abstracted from the complexity of actual experience and if the way in which this is done does not accommodate so basic a feature of human experience as the present moment, so much the worse for physics, one could say. Only an extreme physicalist, supposing that physics is the sum total of all reliable knowledge and understanding, could make this argument the basis of a claim to abolish the reality of the present moment. Moreover, when one moves from physics to cosmology and considers the Universe as a whole, there is indeed a natural meaning of cosmic time (and so of a cosmic 'now'), which is that defined by the frame of reference at rest with respect to the cosmic background radiation. This is the definition that is used in stating that the age of the Universe is 13.7 billion years. It might seem a bit far-fetched to appeal to the overall state of the Universe in order to make a connection with our terrestrial experience of time's passing, but there is another example of this kind of cosmic coupling. A hundred years ago, Ernst Mach pointed out the remarkable fact that inertial frames of reference here on Earth (that is to say, the way that mechanics is most simply described on our planet) are at rest or in uniform motion with respect to the fixed stars. In other words, local physics in our neighbourhood is apparently linked to the state of matter in the Universe as a whole. It appears that there is more interconnectedness in the cosmos than we are naturally inclined to expect.

The supporters of the picture of cosmic history as involving the genuine unfolding of processes of becoming do not only appeal to human intuition, but they can look also to physics for support. The second law of thermodynamics, decreeing the increase of entropy in isolated systems, and the quantum processes of measurement after which a determinate result is obtained for the value of a previously indeterminate property, both define arrows of becoming. The intrinsic unpredictabilities that physics has discovered, both at the subatomic level of quantum theory and at the everyday level of chaos theory, open up the prospect of an open future which is not simply the inevitable consequence of the state of the past. Many believe that here is some clue to how it is that human beings are able to exercise a genuine power of agency in playing their part in bringing about the future. The old classical picture of a clockwork universe of mere mechanism is definitely dead. The physical world is something more subtle, and I believe more supple, than that implausibly rigid account could accommodate.

At this point it is important to note that the metaphysical issue of the nature of time and the metaphysical issue of the nature of causality are independent of each other. The block universe does not necessarily imply a deterministic world, since the pattern of events located within its spacetime continuum could correspond either to tight determinism or to the operation of a variety of causal principles, including agency. Nevertheless, the more open that the pattern of actual causal relationships appears to be, the more natural becomes the picture of a world of true becoming. The acts of truly free agents cannot be known beforehand, nor the results of truly indeterminate processes.

The debate between these two opposing views of the nature of time will continue and it is further explored in Polkinghorne (2005). Ultimately the issue is one that requires for its settling metaphysical decision and argument. It is not a matter for science alone. Physics constrains metaphysics but it does not determine it, much as the foundations of a house constrain what can be built on them but do not determine the exact form of the resulting edifice. That is why the debate between supporters of the atemporal block universe and supporters of the universe of temporal becoming continues unabated. It is a metaphysical dispute.

Before we leave physics, however, there is one further issue to note. The two greatest discoveries of twentieth-century physical science were quantum theory and General Relativity. Yet we enter the twenty-first century with these two theories still imperfectly reconciled with each other. While many place their hope in the development of string theory as being likely to provide the eventual resolution of this embarrassing scientific dilemma, the subject of quantum gravity remains necessarily speculative at the present time. Since General Relativity intimately connects together the natures of space, time and matter, and since the typical consequence of quantum physics is to turn what was classically continuous into a sequence of distinct points, there is a general expectation that quantum gravity will be found to imply some form of a discrete structure of time, understood at the lowest level as its being a succession of separate 'moments'. The size of these basic temporal units would be very small indeed (10^{-43} seconds), which is why we are totally unaware of their existence. Some of the essays in this volume have touched upon this issue.

I am both a scientist and a theologian and I am therefore interested in how theories of temporality may influence theological understandings of how God relates to the time of creation. Once again there are two main contrasting stances that can be taken.[†]

Classical theology, in the great tradition that stretches from Augustine and Boethius (fourth to sixth centuries), through Thomas Aquinas in the later Middle Ages and on through Reformers like John Calvin to the present day, pictured God as existing in eternity, wholly outside time and looking down on the whole history of creation, laid out before the divine eye 'all at once' so to speak. According to classical theology, God does not have foreknowledge of the future, for all events are equally contemporaneous to the atemporal gaze of divinity. In other words, for the classical theologians, God knows creation exactly as the block-universe theorists describe it. Since theologically it is clear that God must know creation in true accordance with its actual nature, this would seem to imply a theological endorsement of the fundamental correctness of the block-universe point of view! This claim is one that is seldom actually made in the theological literature, but it seems an unavoidable implication of classical theology.

In more recent years, however, a different theological approach has been developed, often given the name of Open Theology. This

[†] See Chapter 7 of Polkinghorne (2000) and Chapter 6 of Polkinghorne (2005).

invokes the concept of a much closer relationship of God to the time of creation than classical theology had been disposed to adopt. There is certainly acknowledged to be an eternal dimension of the divine being, since God is not at all pictured as being in thrall to time in the way that creatures necessarily are. Yet to this concept of divine eternity is added the idea of a temporal dimension also, as God engages with the unfolding history of a world of true becoming. Most open theologians picture this divine dimension of temporality as being graciously and freely embraced by God and not imposed on divinity by some outside metaphysical necessity. (An exception to this is the scheme of process theology, based on the metaphysical thinking of A. N. Whitehead.) Such a divine dipolarity of eternity/time not only seems to correspond to the way that the Bible pictures God as being related to human history, but also, if creation is indeed a world of true becoming, then God will surely know it as it actually is, that is to say in its becomingness. God will not just know that events are successive but will know them in accordance with their nature, that is, in their succession. Some theologians, including the present writer, are bold enough to conclude that this implies that even God does not yet know the unformed future. Of course, any theist will believe that God knows all that can be known, but in a world of unfolding becoming the future is not yet there to be known, so current ignorance of its nature would be no divine imperfection.

The nature of time is indeed a challenging issue. We have seen how physics, metaphysics and philosophy, and theology all have a stake in the discussion. A discussion of this kind offers access to the consideration of questions that are of fundamental significance for our understanding of the rich reality within which we live.

BIBLIOGRAPHY

Isham, C. J. and Polkinghorne, J. C. (1993) The debate over the block universe, in *Quantum Cosmology and the Laws of Nature*, ed. R. J. Russell, N. Murphy and C. J. Isham, Vatican Observatory/CTNS, pp. 135–44.

Polkinghorne, J. C. (2000) *Faith, Science and Understanding*, SPCK/Yale University Press.

Polkinghorne, J. C. (2005) *Exploring Reality*, SPCK/Yale University Press.

Index